The Search Institute Series on Developmentally Attentive Community and Society

Series Editor
Peter L. Benson
Search Institute, Minneapolis, MN

For further volumes:
http://www.springer.com/series/6554

Michael J. Nakkula · Karen C. Foster ·
Marc Mannes · Shenita Bolstrom

Building Healthy Communities for Positive Youth Development

Springer

Michael J. Nakkula
University of Pennsylvania
Graduate School of Education
Division of Applied Psychology-Human Development
Walnut St. 3700
19104 Philadelphia Pennsylvania
GSE Room 308
USA
mnakkula@gse.upenn.edu

Karen C. Foster
01810 Andover Massachusetts
USA
kfoster52@comcast.net

Marc Mannes
Search Institute
First Ave. NE., 615
55413 Minneapolis Minnesota
USA
marcm@search-institute.org

Shenita Bolstrom
Medtronic, Inc.
55432-3568 Minneapolis
Minnesota
USA
shenital@gmail.com

ISBN 978-1-4419-5743-6 (hardcover)
e-ISBN 978-1-4419-5744-3
ISBN 978-1-4419-6819-7 (softcover)
DOI 10.1007/978-1-4419-5744-3
Springer New York Dordrecht Heidelberg London

Library of Congress Control Number: 2010926131

Printed on acid-free paper

Springer is part of Springer Science+Business Media (www.springer.com)

Series Preface

It is a great pleasure to offer this volume from Michael J. Nakkula, Karen C. Foster, Marc Mannes, and Shenita Bolstrom as the latest in the Search Institute Series on Developmentally Attentive Community and Society. Its importance to the series and this field of inquiry and practice is readily evident in its title, *Building Healthy Communities for Positive Youth Development.*

Since the early 1990s, Search Institute has invited and encouraged communities of all shapes and sizes to use its framework of Developmental Assets and principles of asset building to create strong, vibrant, and welcoming communities for children and youth. We have operated largely at the grassroots level, encouraging innovation and adaptation around a shared vision, rather than proposing a program or model for replication. We seek to learn as much from the communities as they learn from us. This book offers in-depth case studies of what happened in eight diverse communities that took up our invitation. In them, we see a wide array of strategies and approaches that, on the surface, seem to have little coherence. But, as Nakkula and colleagues found, underlying each of these distinct efforts was a deep commitment to transforming the social norms of community life to more effectively attend to young people's healthy development throughout the first two decades of life.

There have been many ambitious efforts aimed at comprehensive community change on behalf of young people. Many have invested heavily in the transformation of policies, programs, and systems, with greater and lesser degrees of success. Though the communities described in this book have addressed these systems-based issues, their (and our) focus has been weighted toward the informal dynamics, networks, and relationships of community life. You'll see how these communities have invested much of their time and energy in mobilizing everyday community members (youth and adults) to be part of the solution. You'll see formal, systematic planning and action bumping into the more fluid and dynamic processes of movement building. In the process, you'll begin to see the strengths and limitations that diverse approaches offer for community transformation.

Since we first introduced the Developmental Assets in the early 1990s, I have had the opportunity to be in hundreds of communities that have used the asset framework as a guide or springboard for transformation. Drawing from the efforts of eight distinct communities, this book provides the most systematic analysis and interpretation of this approach to community building with and on behalf of youth that is available. As such, it is essential reading for leaders, scholars, and students of community life.

As editor of this series, I am deeply grateful to Mike, Karen, Marc, and Shenita for their thoughtful, rigorous, and tenacious efforts to conduct this research and produce this volume. I am just as grateful to the leaders and residents of the eight communities—representing hundreds of others—who have made it their priority to transform their communities to ensure that all young people in their midst have the opportunities they need to grow up successfully.

Peter L. Benson, Ph.D.
Search Institute
Series Editor

Acknowledgments

Primary thanks for this book go to the youth and adults who so graciously welcomed us into their communities. The many participants of the eight Healthy Communities • Healthy Youth (HC • HY) initiatives featured here inspired us with their stories of growth, challenge, and community partnership. Each initiative made it possible for us to learn critical new lessons for the work ahead.

Special thanks go to the initiative directors and site coordinators who helped us identify and access key contributors in their communities and provided such generous support, despite the additional hours our visits required of their time. Those directors and coordinators include Becky Beauchamp of Traverse Bay Area's GivEm40.24.7; Lisa Stein and Cathy Billings of Lawton/Fort Sill's Community Coalition; Ray Larsen of Orlando's Healthy Community Initiative; Karen Atkinson of St. Louis Park's Children First; Christy McGill and Alecia Hoffman of the Healthy Communities Coalition of Lyon and Storey Counties in Nevada; Cheryl Lynn Higgins of McPherson, Kansas's Tri-County Asset-Building Initiative; Barry Nelson of Moorhead's Healthy Community Initiative; and Chris Tebbin of Portland's Take the Time. We hope we have done your lessons justice and that this book serves as a helpful teaching tool. The constraints of a book like this one, which features multiple approaches from around the country, make it impossible to get any one initiative's story out fully; we have tried, therefore, to extract a variety of lessons across the eight sites, hoping that the sum of the parts creates a representative and informative whole. Our deepest thanks to all of you for allowing us to learn and pass along your knowledge and experience.

We also owe a great debt to the funders who made various aspects of this project and this book possible. These were the Donald W. Reynolds Foundation, Lutheran Brotherhood, Kansas Health Foundation, and the McKnight Foundation, each of which supported the research project. In addition is the Lilly Endowment, which helped make the development of this book possible as part of its support for the Search Institute Series on Developmentally Attentive Community and Society.

Although the four authors worked collectively to lead different aspects of the research design and data collection, analysis, and reporting, many others were integrally involved in the project at key junctures along the way. A special thanks to Dr. Andrew Schneider-Muñoz for helping to conceptualize the project (during his tenure at Search Institute), and for bringing the Search Institute–Harvard University research team together. Andy has a knack for hatching special projects on behalf of youth well-being, and he certainly helped make this one a reality. Mike will never forget the uniquely served orange juice and coffee Andy delivered to his hotel door! We are also grateful to Search Institute's Nancy Tellett-Royce, senior consultant, for generously sharing her vast knowledge of the HC • HY initiative history and evolving network of community partnerships. Sandra Longfellow, manager of the Information Resource Center at Search Institute, was unfailingly helpful in locating the literature we needed.

Research assistants who played key roles in data collection and analysis include Nicole Hintz and Jennifer McGaffey of Search Institute and Anna Cammidge, Linda DiPalma, and Kristina Pinto of Harvard University. A tremendous debt of gratitude is due each of them.

Dr. Peter L. Benson, president of Search Institute, was supportive of this project from its inception and was instrumental not only in making possible the research presented here but also in the development and evolution of the HC • HY initiative itself. Without Peter's efforts, on so many levels, neither this study nor its focus would have been possible.

Many layers of editorial support have gone into this book. Kathryn L. (Kay) Hong, formerly senior projects manager at Search Institute, provided the initial support necessary for framing the study as a readable, informative, and useful book. Kay's pragmatism and aesthetic sensibilities got the book off to a strong start. Copy editor Mary Byers provided invaluable editorial guidance throughout the process, without ever being heavy-handed. Mary's extensive editorial expertise and elegant writing skills are reflected throughout the book. As is so often the case, Eugene C. Roehlkepartain, vice president of Search Institute, was essential to getting the project over the finish line. And, finally, our thanks to the editorial and publication staff at Springer, particularly Judy Jones, Garth Haller, and Jennifer Hadley, for their professional handling of the final details.

Contents

About the Authors

Michael J. Nakkula, Ed.D., is a practice professor of education within the Division of Applied Psychology and Human Development at the University of Pennsylvania's Graduate School of Education, where he serves as the director of a master's program in school and mental health counseling. Nakkula teaches courses on adolescent development and the intersection of counseling, mentoring, and education within urban public schools. Prior to assuming his current faculty position, he was the longtime codirector of Harvard's Risk and Prevention master's program, where he designed and studied a number of initiatives that support the developmental opportunities for low-income youth. For this work, he was named Harvard's initial recipient of the Kargman Junior Chair for Human Development and Urban Education Advancement (1998–2004). Nakkula is the lead author (with Eric Toshalis) of *Understanding Youth: Adolescent Development for Educators* (Harvard Education Press, 2006), and is the coeditor of a special journal issue on the ways in which youth mentoring relationships are organized, assessed, and understood to promote best practices within different settings. He finds particular joy in helping to coach his sons' (Lukas and Sam) youth hockey and baseball teams.

Karen C. Foster, Ed.D., is a former research associate at the Harvard Graduate School of Education, where she enjoyed a longterm collaboration with Michael Nakkula. She trained in child and adolescent development at the former Cambridge Child Guidance Center and in the Behavioral Medicine department, also affiliated with The Cambridge Hospital, in Cambridge MA. Her early research was a study of Boston inner city students' experiences of stress and coping. She has spent the past 4 years as an external evaluator and consultant to the Ohio Board of Regents studying the impact of a statewide initiative to improve the quality of science and mathematics teacher education in the public schools. Her stint in Ohio provided a valuable opportunity to study the unique challenges to higher educational access posed by rural communities. The experience of working on the National Asset-Building Case Study Project was a primary force in shaping her deep interest in research approaches that authentically capture the work of social reforms. She has applied the model developed on the Case Study Project to other areas including immigrant healthcare practices, nutrition reform, service learning, and police profiling.

Marc Mannes, Ph.D., currently leads the Chicago office of Booz Allen Hamilton, a strategy, operations, and technology consulting firm serving government clients. Marc spent a decade as director of applied research at Search Institute, where he oversaw the organization's community research agenda and the extension of the Developmental Assets frameworks to children. His career has focused on the intersections of applied research, policy formulation and implementation, program and product development, organizational and community change, and training. Mannes is coeditor and contributor to *Balancing Family-Centered Services and Child Well-Being: Exploring Issues in Policy, Practice, Theory, and Research* (Columbia University Press, 2001). He also served as a contributing author to *Other People's Kids: Social Expectations and American Adults' Involvement with Children and Adolescents* (2003), also part of the Search Institute Series on Developmentally Attentive Community and Society. Mannes lives north of Chicago in Wilmette, Illinois, with his wife, Karen, and their two daughters, Lilia and Natalie.

Shenita Bolstrom, MS, is a clinical research associate in the Neuromodulation business at Medtronic, Inc., global leader in medical technology, headquartered in Minneapolis, MN. Prior to joining Medtronic, Inc., she was field research coordinator at Search Institute, where she worked on research projects in the areas of community and social change, families and parenting, and youth development programs and practices. Shenita earned her Bachelor of Arts degree in psychology and sociology from the University of Wisconsin–Milwaukee and her master's degree at the University of Wisconsin–Madison. When not at work, she's out training for and completing marathons.

Chapter 1
Introduction

Eight Interpretations of the Developmental Assets Framework

What do Lyon and Storey counties in northern Nevada have in common with Orlando, Florida? What does McPherson, Kansas, share with the Traverse Bay Area of northern Michigan? These communities and hundreds of others across the United States are linked together by a particular commitment to healthy youth development. That commitment is a pioneering effort to interpret and apply Search Institute's Developmental Assets framework according to the individual features of each community and the unique needs and resources of each community's youth. The highly popular framework, which articulates 40 specific aspects of environmental supports and inner resources that children need to live healthy and prosperous lives, had, up until the mid-1990s, primarily been used to help people understand healthy development, but that understanding had not been comprehensively applied to practice. This book is a study of eight communities from across the United States that have adopted the framework, studied its potential for their own communities, and applied it in ways that made the most sense to them given their particular needs and resources.

The eight communities featured here are part of more than 600 towns, cities, and counties that have adopted the Developmental Assets framework as part of the Search Institute–sponsored Healthy Communities • Healthy Youth (HC • HY) initiative. The initiative invites communities to apply the framework locally with little to no prescriptive guidance, as part of an effort to promote innovative approaches to supporting the healthy development of children and youth. Through our study, we learned that some communities found the lack of prescription liberating relative to alternative models of youth development and community change, whereas others felt quite the opposite: They yearned for community models that depicted successful applications of the Developmental Assets framework. Communities on both sides of this issue, and at all points on the spectrum in between, contributed inspiring examples of community-wide youth development efforts for this study. We might think of these examples as *interpretations* of the framework, rather than as models or even approaches, because they are efforts to *translate* a general framework into particular cultural contexts. As with artistic interpretations, each rendering presents novel and,

M.J. Nakkula et al., *Building Healthy Communities for Positive Youth Development*, The Search Institute Series on Developmentally Attentive Community and Society 7, DOI 10.1007/978-1-4419-5744-3_1, © Springer Science+Business Media, LLC 2010

at times, courageous twists on some basic themes. Those community renderings are the subject of this qualitative study.

Specifics of the Developmental Assets Framework and the HC • HY Initiative

Since the early 1990s, Search Institute has synthesized research on the factors leading to healthy development for children and youth. That synthesis led to the Developmental Assets framework (Benson, 1997/2006; Scales & Leffert, 1999/2005), which has guided the institute's ongoing applied research and community development agenda. The framework has also become one of the most visible and widely referenced manifestations of the positive psychology (Larson, 2000; Seligman & Czikszentmihalyi, 2000) and healthy youth development (Pittman, Irby, & Ferber, 2001) movements that have gained ascendancy over the past two decades.

The Developmental Assets framework is most fully articulated in the book *Developmental Assets: A Synthesis of Scientific Research on Adolescent Development* by Scales and Leffert (1999/2005), in which they present the basic model and the literature base contributing to it. The model divides the assets into two primary groups, external and internal, each of which further divides into four asset categories. The external categories consist of Support, Empowerment, Boundaries and Expectations, and Constructive Use of Time. The internal categories are Commitment to Learning, Positive Values, Social Competencies, and Positive Identity (see Display 1.1). The two primary groups and the asset categories serve as organizers of the synthesized interdisciplinary body of basic and applied research focusing on healthy youth development. The distillation of this disparate body of research into a language accessible to developmental scholars, youth practitioners, and policy makers has resulted in the Developmental Assets framework's shifting toward the center of discussion and debate within diverse sectors of the youth-invested community, including regional and national youth development foundations, policy-making arenas, service-providing organizations, and college and university classrooms.

Display 1.1
Search Institute's Framework of 40 Developmental Assets
EXTERNAL ASSETS

Support

1. *Family Support* —Family life provides high levels of love and support.
2. *Positive Family Communication* —Young person and her or his parent(s) communicate positively, and young person is willing to seek parental advice and counsel from parents.

3. *Other Adult Relationships* —Young person receives support from three or more nonparent adults.
4. *Caring Neighborhood* —Young person experiences caring neighbors.
5. *Caring School Climate* —School provides a caring, encouraging environment.
6. *Parent Involvement in Schooling* —Parent(s) are actively involved in helping young person succeed in school.

Empowerment

7. *Community Values Youth* —Young person perceives that adults in the community value youth.
8. *Youth as Resources* —Young people are given useful roles in the community.
9. *Service to Others* —Young person serves in the community one hour or more per week.
10. *Safety* —Young person feels safe at home, at school, and in the neighborhood.

Boundaries and Expectations

11. *Family Boundaries* —Family has clear rules and consequences and monitors the young person's whereabouts.
12. *School Boundaries* —School provides clear rules and consequences.
13. *Neighborhood Boundaries* —Neighbors take responsibility for monitoring young people's behavior.
14. *Adult Role Models* —Parent(s) and other adults model positive, responsible behavior.
15. *Positive Peer Influence* —Young person's best friends model responsible behavior.
16. *High Expectations* —Both parent(s) and teachers encourage the young person to do well.

Constructive Use of Time

17. *Creative Activities* —Young person spends three or more hours per week in lessons or practice in music, theater, or other arts.
18. *Youth Programs* —Young person spends three or more hours per week in sports, clubs, or organizations at school and/or in community organizations.
19. *Religious Community* —Young person spends one or more hours per week in activities in a religious institution.
20. *Time at Home* —Young person is out with friends "with nothing special to do," two or fewer nights per week.

INTERNAL ASSETS

Commitment to Learning

21. *Achievement Motivation* —Young person is motivated to do well in school.
22. *School Engagement* —Young person is actively engaged in learning.
23. *Homework* —Young person reports doing at least one hour of homework every school day.
24. *Bonding to School* —Young person cares about her or his school.
25. *Reading for Pleasure* —Young person reads for pleasure three or more hours per week.

Positive Values

26. *Caring* —Young person places high value on helping other people.
27. *Equality and Social Justice* —Young person places high value on promoting equality and reducing hunger and poverty.
28. *Integrity* —Young person acts on convictions and stands up for her or his beliefs.
29. *Honesty* —Young person "tells the truth even when it is not easy."
30. *Responsibility* —Young person accepts and takes personal responsibility.
31. *Restraint* —Young person believes it is important not to be sexually active or to use alcohol or other drugs.

Social Competencies

32. *Planning and Decision Making* —Young person knows how to plan ahead and make choices.
33. *Interpersonal Competence* —Young person has empathy, sensitivity, and friendship skills.
34. *Cultural Competence* —Young person has knowledge of and comfort with people of different cultural/racial/ethnic backgrounds.
35. *Resistance Skills* —Young person can resist negative peer pressure and dangerous situations.
36. *Peaceful Conflict Resolution* —Young person seeks to resolve conflict nonviolently.

Positive Identity

37. *Personal Power* —Young person feels he or she has control over "things that happen to me."
38. *Self-Esteem* —Young person reports having a high self-esteem.
39. *Sense of Purpose* —Young person reports that "my life has a purpose."

40. *Positive View of Personal Future* —Young person is optimistic about her or his personal future.

The HC • HY initiative provides communities with information on the Developmental Assets, but it does not provide prescriptions for acting on that information. Each community interprets the model as it fits within the local context. Some initiatives work with the full model, paying specific attention to each of the asset categories; others focus on a small number of Developmental Assets or perhaps one asset category. In 1993, Search Institute began working with communities to build structures and processes that strengthen their resources for supporting youth (external assets) as a means of cultivating the inner strengths (internal assets) of young people. In 1996, the HC • HY initiative was formalized, with Search Institute beginning to provide communities with practical resources, technical support, and forums for communicating with one another concerning the challenges of their work. More than 600 communities have become formal partners in the initiative, with countless others using the Developmental Assets framework informally.

As noted, the HC • HY initiative is intentionally nonprescriptive. Nonetheless, the following simple guidelines are offered to help communities get their initial work off the ground. Given the community approach inherent to the initiative, it is recommended that a minimum of three community sectors collaborate in the work. All of the eight communities we studied included more than three sectors, with some choosing to start with a few key sectors and building from there, while others involved many more sectors from the start. Another recommendation is that the local initiatives select at least three core assets on which to focus. This primarily is a start-up recommendation, with acknowledgment that most initiatives will focus on a larger number of assets over time. The purpose of this second recommendation is to help prevent initiatives from feeling overwhelmed by the large number of assets. Starting with a clear focus on a few assets helps ground the initiatives in concrete activities as they build their more comprehensive approaches.

Another recommendation is that local initiatives create their own name and identify a home space within the community to help with larger recognition and identification. All of the communities we studied put a great deal of strategic thought into this "branding" effort, as many of them called it. And, finally, it is highly recommended that a local initiative include youth in leadership roles and as key resources, rather than viewing them as targets of interventions. The nature of youth leadership and more general involvement varied in both form and degree across the eight communities in our study, and ultimately became one of the most interesting findings in our work.

The National Asset-Building Case Study Project

The National Asset-Building Case Study Project (hereafter the Case Study Project) is a qualitative inquiry into the strategies implemented by community stakeholders involved in the HC • HY initiative. Stakeholders come from multiple professional and population sectors to promote what Search Institute has termed "developmentally attentive communities" (Mannes, Lewis, Hintz, Foster, & Nakkula, 2002)—communities that conscientiously attempt to meet the developmental needs of their youth. Professional sectors include, but are not limited to, youth service providers, educators, philanthropists, police officers, politicians, businesspeople, the media, and members of the clergy. Population sectors include youth, parents, neighbors, seniors, and volunteers of varying backgrounds and interests. By definition, the HC • HY initiative involves multiple sectors and is intended to integrate the perspectives of professionals of different backgrounds with those of everyday people invested in the healthy development of the community's youth.

Derived from the mission of the national HC • HY initiative, the following questions guided our study of the eight community interpretations presented here:

1. How do the local initiatives uniquely understand, adopt, implement, and assess the impact of the developmental assets framework?
2. What are the dynamics of change—setbacks and transformations—that occur as communities develop and attempt to sustain their initiatives over time?

The Case Study Project focuses on work sponsored by Search Institute; for this reason, the study was conducted through a partnership with researchers from an external organization (Harvard Graduate School of Education). The external researchers had no affiliation with Search Institute except as investigators on the current and related studies. This "internal–external" team allowed the institute's applied research group to provide information on the goals and objectives of the HC • HY initiative. The external research partners applied this information to shape the research design with Search Institute and to inform the data analysis and dissemination of findings.

Research Design and Methodology: Developing an Ethnographically Informed Perspective

Because the purpose of the study is to understand the processes and meanings of the HC • HY work as it is evolving within particular communities, rather than measuring global and standardized outcomes, we used an ethnographically informed qualitative research design. We were interested in understanding asset-based change as it occurs in the boardrooms and kitchens, law offices and classrooms, and community centers and skate parks that make up the contexts referenced in this study. In exploring these various contexts, we were interested in the perspectives of law enforcement officers and owners of small businesses, of middle school youth and the elderly, and of community leaders and everyday community members. To capture these multiple perspectives, we conducted individual interviews and focus groups with initiative participants, as well as documenting discussions from staff meetings and informal conversations by means of detailed field notes. Although we were not participant observers in the classic ethnographic sense, in that we did not participate in the work or live in the community, we conducted field visits during our two interviewing trips to gain a sense of the lived experience of the local initiative members.

To maximize the different perspectives of our internal–external team, data collection trips and all phases of analysis included researchers from both Search Institute and Harvard. In addition, we used multiple modes of data collection to address the biases of any one method. The data include individual interview, focus group, and staff meeting transcripts; observational field notes and memos; and review of print materials and artifacts (objects or symbols, such as art, logos, and awards, that represent the meaning of the initiative or local culture under study). The research design included data collection trips at two points in time to offset biases inherent in the timing of the first visit, to share our preliminary interpretations with the communities and have them challenged, and to collect follow-up information. The two visits were approximately 6 months apart, allowing preliminary analysis of Time 1 data to inform Time 2 follow-up interviews.

Site Selection

The eight HC • HY initiatives were selected according to the following criteria: length of existence as a formal HC • HY initiative (minimum of 3 years); geographic region (seeking a range); population demographics (urban, suburban, and rural); and type of initiative (for example, primarily school based versus more broadly community based). As evidenced in the initiative descriptions found in Display 1.2, even though our selection of sites was small for purposes of focusing in depth, we were able to meet our sampling criteria. With a sample of only eight sites, and given the in-depth, site-specific focus of our data collection and analysis procedures, our intent is not to generalize these findings to other HC • HY initiatives but rather to uncover models that may inform our understanding of the divergent and perhaps common pathways by which community change on behalf of healthy child and youth development is pursued and ideally comes about through these initiatives.

Display 1.2
The eight healthy communities • healthy youth initiatives

Community	Location	Urban/ suburban/ rural	Key features
Traverse City, Michigan	Northern Midwest	Rural/small town	Heavily school-based initially; strong funder and evaluation relationships
Moorhead, Minnesota	Upper Midwest	Small city	Focus on neighborhood pride, particularly poor and minority neighborhoods; strong parks and recreation and law enforcement focus
Portland, Oregon	Northwest	Large city	Emphasis on positive media representation of and by youth; use of social organizing theories
Orlando, Florida	Southeast	Large city	Community stabilization due to highly transient population; strong Latino youth leadership orientation
St. Louis Park, Minnesota	Midwest	Suburban	Self-defined as a way of being or "calling," rather than program or even initiative; highly invested community leadership
Lawton/Fort Sills, Oklahoma	Southern Midwest	Small town military base	Focus on high risks facing local youth; innovative blending of prevention and Developmental Assets
McPherson, Kansas	Midwest	Small town	Focus on all people as leaders; rooted in Chamber of Commerce
Northern Nevada	West Southwest	Rural, county based	Wide expanse across five counties; blend of prevention and Developmental Assets

Qualitative Instrumentation

For each of the eight sites, data collection began with an initiative staff meeting, in which the leadership teams and key partners discussed the structure of their initiative, the major tasks undertaken, and the primary sectors involved. These meetings were audiotaped and transcribed for subsequent analysis, and extensive field notes were taken. An average of 10 participants attended the staff meetings across the eight sites. Staff meetings concluded with confirmation of plans for the focus groups. Within each site, one focus group was conducted during the first visit. It included key representatives from each of the community sectors involved in the initiative, including the funding sector (foundation program officers), formal leadership from the initiative, and selected youth who had been particularly involved. Focus groups ranged from 8 to 12 participants across the eight sites. A similar semistructured interview protocol was followed across the focus groups. It focused on the history of the initiative, current structure and activities, the nature of representation (who is most and least involved, and why), and key challenges. Following the focus group interviews, members from the focus group were selected for in-depth individual interviews, along with key initiative participants who did not participate in the focus groups. Twelve individual interviews, on average, were conducted per site, with the protocols focused on the specific nature of the participants' involvement and on their unique perspectives on the work. Following preliminary analysis, a second visit was made in which results were presented to the community participants for their feedback and to help clarify ambiguous meanings. The results were presented in both group and individual meetings; all group meetings were audiotaped and transcribed to inform the final analysis and interpretation of the data.

Analysis: Deriving Data-Driven Interpretations

An inductive grounded theory approach was applied to the data in a manner that allowed themes to emerge without theoretically prescriptive guidelines. Grounded theory analysis is a widely used technique that strives to make meaning of complex data from the ground up; that is, it focuses thoroughly on respondents' understanding of things rather than using predetermined theoretical categories to look for particular traits or characteristics. Search Institute and Harvard coders of the interview transcripts first applied open-ended, largely descriptive codes to the data, followed by a more complex interpretive clustering of themes. The specific analytic steps are as follows.

We began the analysis with *open coding*, which has been defined by Strauss and Corbin (1990) as a process of identifying and labeling largely descriptive units of meaning that are close to the text (interview transcript or observational field notes). Although open coding requires a degree of abstract interpretation in the naming of descriptive meaning units, the process of abstraction is rather minimal given the effort to stay close to the narrative data. Our open-coding procedure consisted of several steps. First, a *code construction* step was undertaken for Time 1 analyses, with two internal (Search Institute) and two external (Harvard) members of the research team reading the community focus group transcripts from each of the eight HC • HY initiatives. Based on this reading, we identified approximately 66 open codes that were retained in the analytic process. Second, these open codes, derived from approximately 500 pages of focus group and interview transcript text, were then labeled, defined, and organized into a codebook according to the procedures described by Boyatzis (1998).

Next, a *code application* phase ensued with two members of the team systematically applying the codebook to each of the eight Time 1 focus group transcripts. This phase resulted in an expansion of the codebook to 80 consensually agreed upon open codes. A first step in reliability or coder-consistency checking followed the initial code application process. The internal and external team members who applied the codebook to a given transcript compared results. Where differences occurred, the two coders and one or two additional team members debated and consensually determined the proper code application. Because differences in applying the codes were quite extensive in this initial phase of code application, all open code applications were consensually determined; as such, this phase of our methodology has been named *consensus coding*. All Time 1 focus group interviews were consensus coded.

Our next step consisted of clustering the open codes into what Strauss and Corbin (1990) defined as axial codes in their explication of grounded theory analysis. Axial codes represent the thematically grouped clusters of open codes. As a first step in axial coding, all of the open codes were divided into preliminary clusters, based on analytic memos constructed throughout the open-coding process and on discussion and debate among team members following the completion of the open-coding process. The two external team members led this process, reviewing, debating, and revising the preliminary clustering. We then settled on labels that adequately reflected the meaning of the axial code clusters and created definitions for them.

Although we did not set out to distill the data into one *core code*, which, according to Strauss and Corbin (1990), depicts the essence of the research story, we ultimately identified a core code that succinctly depicted what we had learned. In Chapter 2, we provide an overview of our qualitative findings, from the open-coding process right through to the core code. The overall findings presented in Chapter 2 are then applied to each of the eight community initiatives featured here.

Unfinished Collaborations, Dynamic Processes

It is important to note that the eight community interpretations that follow represent development through a particular point in time. Each of the models presented is fluid by virtue of the dynamic nature of community collaboration. We present the community interpretations as we understood them at the time of data collection and analysis. Although we completed the majority of data collection in 2002, we—particularly representatives from Search Institute—continued to communicate with leaders from the initiatives as their efforts changed shape, formally ended, and at times were dramatically reconfigured. Rather than continuing to add the evolving nature of each community's development, we have chosen to represent the community interpretations at the point at which we had our most substantial data and were able to complete our most thorough comparative analyses. (Brief descriptive postscripts to each case provide updates on how each effort has changed, strengthened, or faded since data were collected.) As a result, these interpretations should be read as exemplars of processes for community change, rather than as completed or static models. It is through the processes presented, we believe, that each of the eight local initiatives has so much to teach about community development and collaboration for the promotion of healthy youth development.

Chapter 2
Transformation, Affirmation, and Blended Models

The central purpose of our study was to explore the range of approaches used by the eight Healthy Communities • Healthy Youth (HC • HY) initiatives in their efforts to implement the Developmental Assets framework. We discovered a fascinating array of strategies used across the initiatives, including approaches that were common to the eight community sites and others that were unique to one initiative or another. The overarching story emerging from our analysis is one of *transformation*, *affirmation*, and *blended* (or *braided*) models. Transformation refers to the dramatic changes in community life and functioning that are required to effectively promote healthy youth development. We heard hopeful stories of perceived transformation in action and concerned stories about how far a community needs to move in order to achieve the transformation necessary to accomplish an initiative's aims. We also heard repeated examples of affirmation, in which HC • HY participants described the asset framework as affirming their already held beliefs and behaviors regarding what was important and essential for healthy youth development. But affirming one's actions and point of view is one thing; turning that affirmation into a cohesive community-wide effort is quite another. The stories of affirmation included here center on efforts to translate convictions into sustainable processes shared by progressively larger proportions of the community. Finally, blended models represent efforts to connect the Developmental Assets with existing prevention and intervention approaches by blending or braiding particular aspects of different frameworks. A key question that stems from this finding is this: How can we use a Developmental Assets approach in different ways in response to different levels of youth need? In other words, how do we adapt and customize a Developmental Assets approach and integrate it with prevention and intervention efforts when youth clearly need all three responses?

Whether we examine intentional strategies for transforming community attitudes and behavior, actions to turn affirmation into tangible processes, or the strategic planning required for integrating asset-based efforts with prevention and intervention approaches, there are particular leadership structures, processes, and activities that are important, if not essential, to sustainable asset-building efforts. In some cases, these leadership constructs represent process shifts (*how* lead-

M.J. Nakkula et al., *Building Healthy Communities for Positive Youth Development*, The Search Institute Series on Developmentally Attentive Community and Society 7, DOI 10.1007/978-1-4419-5744-3_2, © Springer Science+Business Media, LLC 2010

ership is enacted); in other cases, they represent results of leadership actions (*what* has been changed, for example: reaching larger numbers of youth through asset-building activities; prompting change in school practices). In addition, we have found multiple forms of collaboration that yield a range of rich asset-building activities that could not have emerged through more traditional, less collaborative approaches. In other words, the collaborative processes have led to enhanced resource development and a more expansive range of asset-building activities.

We should note that our findings primarily address processes (strategies and methods) for applying the Developmental Assets framework. The tangible outcomes from this work can only emerge over time, as communities sustain and continuously build upon their efforts. Perhaps the most common theme to run throughout the eight communities is that this work requires "faith in the long haul." Explicating the nature and importance of sustained commitment to community-based Developmental Assets work over the long haul is, we believe, the central contribution of our study.

Organizing Themes and Concepts

Our interpretive analysis uncovered 82 clearly distinguishable basic *codes* or *essential concepts*, which we then grouped into 11 higher-level *thematic clusters*. In some instances, the essential concepts fit into more than one thematic cluster; in such cases, we included them in the clusters to which they contributed most strongly. The next step in our analysis entailed dividing the thematic clusters into two *primary dimensions* that summarize the most global themes emerging from the analysis. See Fig. 2.1 for a display of the overall coding structure.

The first primary dimension, the New Norm, represents key characteristics related to the new socially sanctioned ways of interacting with and supporting youth that the HC • HY initiatives are attempting to accomplish through their work. The second dimension, Catalytic Context, captures the complex dynamic processes that the initiatives undertake in their efforts to produce the new norms. In the final step of the analysis the two primary dimensions are integrated into a single *core theme* around which the entire analysis revolves. We labeled this theme "Catalyzing the Transformation of Community Norms," in an effort to capture the interactive essence of the two primary dimensions (cohesive theoretical rationale) and in an attempt to communicate a theme that feels honest to the intentions of the HC • HY respondents in our study (grounded empirical rationale). In the sections that follow we present the findings in reverse order, beginning with the core theme and building the justification of it through a presentation of the concepts, clusters, and dimensions from which it was created.

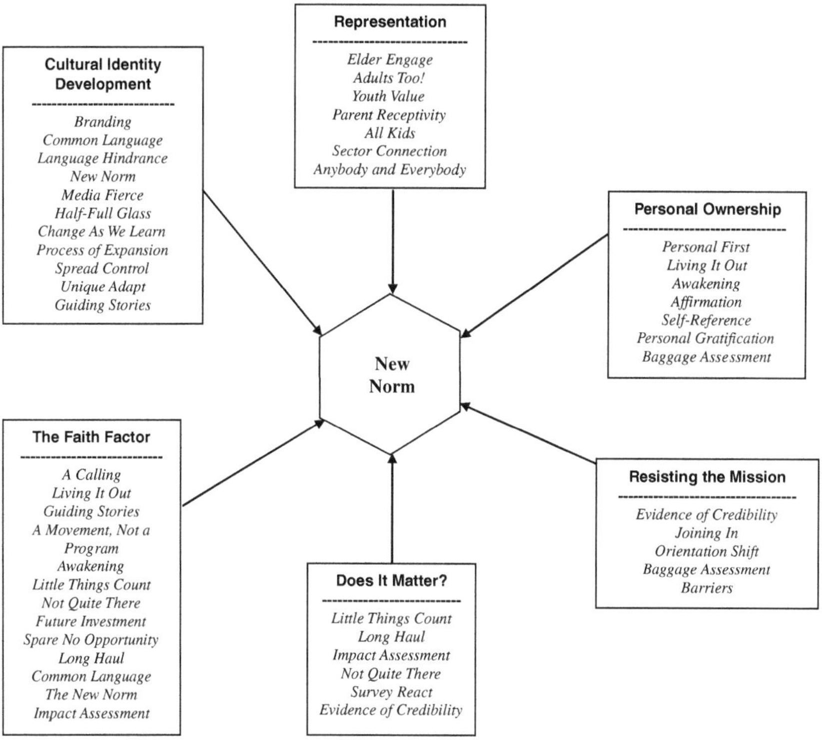

Fig. 2.1 The new norm and its thematic clusters

Catalyzing the Transformation of Community Norms: The Core Theme

Through the process of collecting our data, building our codebook of fairly descriptive, close-to-the-data codes, developing the more interpretive clusters, and repeatedly stepping back to review, assess, and debate the accuracy of our analytic outcomes, we progressively gained a clearer picture of the central story emerging from our study: the participants' successes, challenges, and often struggles to transform in fundamental ways how their communities perceive youth, which, in turn, support their development. We labeled this core code with careful attention paid to each key word. "Catalyzing" was selected because all eight initiatives were fully immersed in the process of bringing sectors of their community together to ignite the initiatives' work; that is, they were actively involved in the "mobilization" and early "action" phases of community change, as described by Mannes and colleagues (Mannes, Lewis, Hintz, Foster, & Nakkula, 2002). Even though the initiatives had all been operating for several years at the time of our data collection, and although there were clear differences in the status of their development, they were all working particularly hard to catalyze growth, whether it was expansion of the initiative itself or start-up activities within particular sectors of the community. For example, "sustaining" the work of the initiatives was a less-pronounced theme than catalyzing that work, perhaps because of where the initiatives were in their development. We imagine that "sustaining normative transformation" will be an important theme as the initiatives continue to mature.

The word "transformation" was selected with cautious consideration. It risks smacking of grandiosity and idealistic ambition, but as we analyzed the responses to our interviews, and reviewed our analyses with initiative participants, transformation, complete with the weighty expectations it implies, accurately represents what we learned. The struggles the initiatives face in carrying out their work can be linked to the lofty goal of "transforming community norms." If communities were readily receptive to the change efforts undertaken by the HC • HY initiatives, there arguably would be little need for their existence. The fact that they do exist, and are organized in connection with what Search Institute defines as a "national movement," implies that the purpose of these initiatives is indeed grand and aptly termed "transformational."

"Community norms" are emphasized in this core theme because the initiatives are not simply attempting to catalyze growth or change in specific institutions; they are aggressively working to transform deeply engrained community patterns of thought and behavior regarding the way youth are perceived and treated. Although HC • HY efforts are directed primarily toward changes at the larger community level, they also target key professional and population sectors of the community. References are made, for example, to transforming the ways in which media depict youth, teachers interact with them, service providers treat them, and police officers engage with them. Respondents communicated a keen recognition that the focus of their work was not on program development or specific types of reform but, rather, on countering the pervasive negative portrayal and problem-based treatment of young people, and on developing effective strategies for implementing an alternative vision grounded in hope, possibility, and developmental well-being.

With this description of our core code in mind, we move toward representing a more differentiated view of our findings. We begin with the primary dimension, the New Norm, including its six thematic clusters, and follow that with our presentation of Catalytic Context and its five clusters.

The New Norm

Ultimately, what are the HC • HY initiatives trying to accomplish through their community change efforts? Although the initiatives are linked through the larger mission of Search Institute's community change movement, each works toward this global mission in its own way, according to the history, interests, needs, and resources of each individual community. The core of each effort, however, came across as quite similar: the development of a healthier way of seeing or understanding youth, resulting in more proactive opportunities for interacting with them and fostering their development. We named this core component of the mission the New Norm, which we define as the ongoing need to foster and reinforce attitudes and practices oriented toward a view of youth as fundamentally healthy and capable of making constructive contributions to the quality of community life, rather than as inherently problematic and in need of community management. This new perspective views youth as critical resources who need ample supports and opportunities to fully develop rather than as problems who cost the community its resources. The aim of the initiatives is to establish this alternative view as normative at the community and even societal levels. Virtually no one was under the impression that establishing the New Norm would be easy, and many respondents expressed concern that the barriers are substantial enough to undermine even their best efforts. In this vein, it is important to reiterate the repeated assertions of respondents in our study that promoting the shift in question is an ongoing task, often accompanied by resistance, setbacks, and years of deficit-based thinking.

The New Norm also emerged as one of our two primary dimensions because it intersects with so much of what the initiatives described as specific components of their goals, not just their larger mission or philosophical aims. For example, a great deal of emphasis was placed on the sector-specific relationships within the initiatives, with discussion centering on the particular asset-based work needing to be done in each sector, and the shifts in perspective and working methods needed to bring about a new way of seeing and doing the work of youth development within those sectors. A primary sector for all initiatives is the funding or philanthropic arena. Two common challenges for that sector were repeatedly noted, such as, shifting funders' priorities away from providing discrete "services for" to promoting ongoing "activities with" youth, and having funders join the initiative as active participants and shapers of the mission rather than standing back as objective providers of revenue in order to be fair to all funding applicants. The intent of the initiatives was neither to dissuade foundations and other funders from supporting services and programs altogether, nor to co-opt them into an insider status that might limit opportunities for noninitiative funding applicants. It was, rather, to reorient the funding community's perspective toward an alternative view of youth and youth development—one less rooted in short-term problem remediation and more aligned with long-term investments in healthy development through sustainable cross-sector partnerships.

Each sector has its own challenges in cultivating the New Norm, and there are additional challenges to communicating and working across sectors, with new strategies needing to be developed for that task as well. One of our basic codes or essential concepts contributing to the New Norm is *Common Language*, which refers to the use of the Developmental Assets framework as a linguistic aid in communicating about youth development within and across sectors. The asset language can serve as an organizer for bringing sectors together in their common work of supporting youth; in short, the assets and the larger framework for describing them often constitute a starting point for cross-sector work, toward which each sector can make its unique contributions. On the other hand, the asset-based language sometimes creates barriers to understanding and working together, a finding we coded *Language Hindrance.* Not everyone is comfortable thinking about developmental strengths or even youth themselves as "assets," given the obvious linguistic connection to investment, banking, and capitalism in general. Some participants in our study said the language was simply confusing because it was a new way of talking about development and youth. The dialectic of common language and language hindrance creates an important tension in the development of the new community norm.

The New Norm is composed of the following six thematic clusters: *Representation, Cultural Identity Development, The Faith Factor, Resisting the Mission, Personal Ownership,* and *Does It Matter?* As noted earlier, each cluster is a composite of several essential concepts, or fairly descriptive codes, many of which contribute to more than one thematic cluster. Each cluster is intended to differentiate one key meaning composite from another under the umbrella of the New Norm, thereby highlighting the most prominent aspects of this primary dimension (Table 2.1).

Table 2.1 Thematic clusters comprising the new norm

Thematic clusters	Definitions
Representation	The demographic and political composition of the initiative. Intended to capture the populations (youth and adult) and community sectors involved in the initiatives.
Cultural Identity Development	The community-specific nature of the local initiative. Emphasizes the community characteristics, youth development practices, and rituals that make each initiative unique.
The Faith Factor	Depicts the faith required to pursue and persist in community-wide youth development practices, given the degree of challenge involved, and the lack of systematic outcome research to date.
Resisting the Mission	Captures the many reasons for resisting the mission of the initiatives, including commitments to alternative models of youth development, disagreements with leadership philosophies and practices, and the need to see results before committing.
Personal Ownership	One must apply the positive shift in attitude toward youth to one's own views and behavior before promoting it to others. Suggests that the larger work of the initiatives can only move forward when participants "live out" the approach in their own lives first.
Does It Matter?	The various ways of assessing whether the initiatives are having an impact or bringing about meaningful change.

Representation

Central to the New Norm is a cluster of codes or concepts that reflects the political dynamics and demographics of the initiative. We named this thematic cluster Representation to capture its descriptive aspects—who is involved—and its political essence: an emphasis on those commonly marginalized or uninvolved in youth development work. Representation asks these: Who is represented in the new norm? How widespread is the norm? Who has a voice in shaping it and in carrying it out? Is the norm applicable to all children and youth, and are the responsibilities for realizing this norm shared by all sectors of the community?

The concept *Anybody and Everybody* captures the "everyperson" component of Representation. It is defined as not needing to be in a position of authority, of a particular status, or in possession of special skills or abilities to build or share assets. Focus group respondents discussed the importance of "nonspecialists" coming to view themselves as essential to healthy youth development, and argued that the Developmental Assets framework is fundamentally differentiated from specialized programs by virtue of its emphasis on everyone's role in the development of the community's youth.

Everyone's Responsibility is a companion concept to *Anybody and Everybody*. It captures the ethical importance and pragmatic necessity of garnering broad-based commitment to the work. Whereas *Anybody and Everybody* suggests that all members of a community are important to and capable of carrying out Developmental Assets work, *Everyone's Responsibility* argues that involvement is more than an open invitation: It is both a moral imperative and a pragmatic requirement undergirding the very possibility of building the initiative into a representatively normative and sustainable way of life. The Orlando initiative, for example, emphasizes the importance of involving "white guys over 50," a demographic, they argue, that typically is left out of grassroots attempts to make change in the community, largely because they are perceived as the wealthy status quo likely to be resistant because change would be required of them. Reinforcement of that stereotype, they claim, reduces the power of efforts to nurture a broad-based, community-wide norm for understanding youth and supporting their optimal development (Table 2.2).

Sector Connection and *All Kids* capture two very different and equally important components of Representation. *Sector Connection* refers to the myriad ways that each sector of the community can and does play a role in and contributes to the advancement of the initiatives. Because, by definition, HC • HY initiatives involve a minimum of three community sectors in their work, the nature of multisector involvement is an essential ingredient in the different applications of the Developmental Assets model. How does the police department work with the initiative? The religious community? Parents? In a community-wide initiative, each sector is as important as the others and has an essential contribution to make. Nonetheless, focus group respondents consistently stated the difficulty of involving all sectors simultaneously, and at equal levels of investment. Typically, one or more sectors play lead roles in the work, and the initiative's identity and representativeness build from there. A key challenge discussed in the focus groups centered on whether sector-specific strategies might be developed as guidelines or models for the Developmental Assets movement, given that sector contributions can be quite distinct.

All Kids is an interesting twist on a theme common in youth development work: not overtargeting a specific group to the exclusion of others. The twist is that in juxtaposition to deficit-based approaches to youth work, approaches that typically focus

Table 2.2 The thematic cluster representation and its essential concepts

Representation: The demographic and political composition of the initiative. Intended to capture the populations (youth and adult) and community sectors involved in the initiatives.

Essential concepts/codes	Definitions
Adults Too!	References to adults needing developmental assets and related forms of support, just as youth do, or to adults needing support in learning how to pass along the assets to youth because they may not have had the assets themselves.
All Kids	The focus of assets should be on all youth, not just those who are troubled or have the greatest amount of need; also includes encouraging involvement, engagement, and leadership among all youth, not just the "star kids."
Anybody and Everybody	One does not have to be in a position of authority, of a particular status, or have special abilities to build or share assets or personal stories.
Connectivity	Bringing together divergent sectors of people in support of the common mission of the initiative.
Elder Engage	Senior citizens referenced as an important sector of the community that needs to be engaged in the initiative.
Everyone's Responsibility	Everyone is responsible for playing a role in promoting healthy youth development and building a positive community.
Key Bridges	Critical connections across sectors facilitated by a key individual, organization, or group.
Parent Receptivity	Parents' response to the initiative or asset framework.
Sector Connection	Specific ways in which sectors connect with the initiative and each other.
Youth Value	A demonstrated genuineness by members of the initiative, community, or different sectors toward youth, which is marked by engaging, empowering, listening, valuing, soliciting, and respecting youth's opinions, ideas, feedback, or efforts.

on youth deemed to be particularly involved in high-risk behavior to the exclusion of others whose needs may be less recognizable, asset-based work runs the risk of over-involving youth leaders. The challenge implied in *All Kids* is to secure broad youth participation in the initiatives, and to do so by involving them as active contributors to the initiatives, not just as passive recipients of asset-based activities.

In Wave II of our study, we heard an interesting twist on the *All Kids* theme. Through what we came to call "blended models," or the blending and braiding of approaches to asset-based work, in which the Developmental Assets framework was integrated with prevention or intervention efforts, we gained a clearer understanding of how children and youth might be targeted and involved in different ways, based on the nature of their particular needs and strengths. Youth who are in need of intervention for school truancy or a mental health challenge, for example, might be hard to reach through a purely asset-based approach. Integrating targeted prevention or treatment strategies with an asset-based orientation, however, might enhance the likelihood that all youth will be the focus of a community's efforts, even if some youth are viewed as needing specialized types of support. In the Wave II case study chapters that follow, we present specific examples of such blending and braiding approaches.

Cultural Identity Development

This thematic cluster merges two interrelated categories of responses presented by the study participants: the cultural center of the initiatives and the development of a unique initiative identity. Because the national Developmental Assets movement is applied within the particularities of each local community, the work has a different feel and look wherever it is conducted. We refer to this localized nature and meaning of the work—the rituals, emerging traditions, and regional particularities that influence each community's HC • HY strategies—as the cultural identity of the initiatives. As with individual identity, the coalescing of a core sense of one's work, community, or cultural center is anything but static. To capture the evolving cultural identity of the local initiatives, we emphasized "development" in the labeling of this cluster.

Struggle and the attempt to transform struggle into strength and wisdom are inherent to virtually all models of identity development. Cultural Identity Development includes the critical role of struggle in each community's effort to adapt and apply the principles and ideas of the national movement at the local level. The code *Unique Adapt* captures the community's efforts to make this cultural adaptation. *Change As We Learn* makes explicit that the process of adaptation is dynamic, resulting in changes in initiative identity, strategic focus, and the nature of its core activities, including, in some cases, the shift to blended and braided approaches as described earlier. The process of adaptation and change is, as in all cultures, heavily guided by language. *Common Language* and *Language Hindrance* are companion concepts—two sides of the same coin—that, on one hand, reflect the importance of having a shared asset-based language to help build and communicate a common understanding of the work within and across sectors and, on the other hand, depict how this language can potentially alienate certain segments of youth and adults by virtue of its tone (investment language) and specialized meaning. For example, school personnel speak in an educational dialect common to their training, just as youth service providers are trained in their profession's organizing concepts. The Developmental Assets language attempts to provide a language synthesis that is simple and straightforward and that allows the entire community to speak in a common tongue. Even with that intent, however, there is a need to learn new concepts, which can be challenging for those professionals immersed in their own professional dialect.

Branding represents a particular form of communication. Defined as a name and visual representation that identify the initiative, typically for the purpose of enhancing its visibility to the public, *Branding* serves as a marker of the initiative's place within the larger community, and as an attempt to enhance its scope and representation. Portland's Take the Time initiative highlights a focus on relationship building between youth and adults that permeates the activities central to its work. By vigorously promoting its Take the Time message through the media, the Portland group attempted to nurture a culture of relationships that would leave an imprint on those who came in contact with it. That imprint communicated the message that each individual's time and effort are essential to the healthy development of Portland's youth—time and effort specifically invested in relationship building with young people, getting to know them more personally through the obvious roles of parents and teachers, and also through the less obvious ones such as merchants, police officers, and advertisers (Table 2.3).

Table 2.3 Cultural identity development and its essential concepts

Cultural identity development: The community-specific nature of the local initiative. Emphasizes the community characteristics, youth development practices, and rituals that make each initiative unique.

Essential concepts/codes	Definitions
Branding	A name and trademark, such as a logo, symbol, or slogan, that identifies the initiative and enhances its visibility.
Change as We Learn	Initiatives learn from their successes, challenges, and mistakes, which results in making a shift in direction or action.
Common Language	The asset framework provides a commonly understood language for positive youth development, helping people to better understand and feel connected to the work of the initiative.
Guiding Stories	Sharing of common or symbolic stories that represent and organize the work, experiences, and history of an initiative.
Half-Full Glass	Recognition of wellness, strength, or positive resources within youth, the community, or the initiative in contrast to references of deficit models or "half-empty" attitudes and behaviors.
Language Hindrance	The asset language is hard to grasp by some individuals and sectors, potentially rendering the initiative less effective.
Media Fierce	Strong focus on using the media positively to alter the usual negative portrayals of youth.
New Norm	The vision and goal of working toward a community in which asset building is a natural part of day-to-day life.
Processes of Expansion	Ways in which the initiative develops through intentional strategies or unintentional events or circumstances.
Spread Control	The need for choosing certain directions rather than others, making choices that eliminate some directions and prioritize others.
Unique Adapt	The need for the framework to be adapted to the uniqueness of different communities, sectors, or groups of people.

If *Branding* serves to represent the cultural identity of the initiative outward to the larger community, *Guiding Stories* help build cultural coherence internally and are critical to symbolizing the cultural and developmental history of the initiatives. The term *Guiding Stories* refers to "legends of the work" that have evolved internally and that capture the essence of what is possible for the initiative to accomplish. In the Traverse Bay Area, for example, stories of the 19 school district superintendents all buying into the asset framework symbolize what is possible in other sectors.

These "superintendent buy-in stories" have become transcendent in some respects, depicting possibility not only within the Traverse Bay community but also across other initiatives. In fact, the stories are so pervasive and compelling that they risk overrepresenting Traverse Bay's GivEm 40 as predominantly a "school-based" initiative, potentially minimizing the important work evolving in other sectors. *Guiding Stories* also include common readings or songs shared by the initiative. For example, Malcolm Gladwell's popular book *The Tipping Point: How Little Things Can Make a Big Difference* (2000) helps inspire the work of many HC • HY initiatives within and beyond this study because of its focus on how small, everyday efforts consistently applied can mushroom into transformational shifts in the ways we see and do things.

The Faith Factor

Belief and inspiration are central to this cluster. Because the work is commonly referred to as *A Movement, Not a Program*, sustained participation requires faith that this movement is on the right track and will lead to the desired ends. The concept *Future Investment* captures this sentiment in suggesting that the current work, which might not show tangible results immediately, is an investment aimed at securing a deeper, longer-term benefit: a future healthier community nurturing healthier youth. Unlike more targeted programs focusing on specific issues such as substance abuse prevention or social skills development, the "mission" of the Developmental Assets movement reaches much further. The mission of the initiatives is to alter the community's, and ultimately society's, view of youth from a deficit- to a strength-based orientation (*Orientation Shift*), accompanied by the goal of promoting large-scale changes in service delivery from prevention and remediation (problem focused) to asset building at the community and individual levels (strength or health focused). A mission of this magnitude would seem to require inspired participation (*A Calling*) and faith that the ultimate outcomes, although taking time to achieve, are worth the investment (Table 2.4).

Awakening represents the "ah-ha" experience described by participants as a clear recognition that one has been viewing youth in a particularly negative way, leading to the awareness that this perspective is unhealthy and needs to change. Such awakenings were reportedly inspired by a host of experiences, including attendance at a Search Institute conference or presentation, reactions to one of Search Institute's Developmental Assets surveys, personal and community crises, and stories in the media. Similar to conversion experiences inherent to other faith-based movements, the awakening is but a first step in a long-term process. Respondents talked about the importance of internalizing the asset framework and applying it to one's own life (*Owning It First*), including taking careful stock of one's own deficit-based views of youth (*Baggage Assessment*), in order to apply the work effectively with others. They talked with conviction about living out the framework in their everyday activities—with their own families, at their places of employment, and with other community members (*Living It Out*). And they were clear that the success of the asset movement would only be experienced over the *Long Haul.*

Table 2.4 The faith factor and its essential concepts

The faith factor: Depicts the faith required to pursue and persist in community-wide youth development practices, given the degree of challenge involved, and the lack of systematic outcome research to date.

Essential concepts/ codes	Definitions
A Calling	Feeling personally drawn to a special or "larger than everyday life" mission of asset-based work.
Awakening	Recognition of negative biases held toward youth, which, when recognized, results in viewing youth more constructively.
Future Investment	Recognizing the importance of investing in children and youth in the present as a means of building a better future.
Impact Assessment	Assessing whether the commitment is making an appreciable difference in the community or society in general.
Little Things Count	Even the smallest efforts when consistently applied make an important contribution to healthy youth development.
Living It Out	Illustrations of day-to-day actions in one's personal or professional life that depict a clear commitment to the initiative.
A Movement, Not a Program	Refers to the initiative being more of a social movement than a program, and therefore harder to define and clearly assess.
Not Quite There	Sustained commitment upon recognizing that the initiative is currently falling short of desired goals and expectations.
Spare No Opportunity	Initiative participants utilize all contexts, including the focus groups, to "spread the word" regarding their work.

Resisting the Mission

Faith in the asset framework and the work of the HC • HY communities is met with various forms of resistance. Perhaps the core of resistance is skepticism that "the movement" is, as respondents from the GivEm 40 initiative in the Traverse Bay Area put it, "just another program with a shelf life." *Joining In* is a code created to capture the multitude of reasons for people and sectors joining the work of the initiative at particular points in time. We learned that *Joining In* often requires working through skepticism and resistance. A clear example of this involves the asset-based work of police officers. As the chief of police in the Traverse Bay initiative, GivEM 40, so vividly put it:

> The hard part of dealing with that [asset framework] is getting [it] integrated within our services ... how do you deal [in an asset-based way] with somebody when you're putting the snatch on him?...I mean when you take him into custody. And [the asset work is] a different approach than ... what we're used to ... Before it was very up front, very, you know, this is the way it happens. [Now] we have officers in our schools who have gone through the asset-building [training] and [are] trying to utilize those particular, different assets in their everyday day-to-day operation. Not only with youth ...

Police officers, educators, and human service providers alike discussed the role of taking stock (*Baggage Assessment*) of their entrenched views of youth that often accompany long-term participation in disciplinary or pathology-based work.

Activities such as punishing "law breakers," managing "acting-out behavior," and treating "delinquents" shape not only one's perspective but also the very language providers internalize for describing their work with young people. For these professionals, joining a movement focused on a very different view of youth requires something other than faith; according to many of our respondents, it requires *Evidence of Credibility*. We found that credibility, and evidence of it, came in many forms. For those seeking traditional quantitative evidence, Search Institute's survey-based results were compelling, specifically findings related to the inverse correlation between the presence of assets and reduction of high-risk behavior. For others, evidence took the form of successful asset-based activities in which they participated, or of testimony from respected colleagues, particularly key leaders within their organizations or professions. But credibility based on logic and the modeling of others seems to go only so far. Longer-term, sustained countering of resistance ultimately requires results, some tangible form of impact that allows participants to assess whether their faith or at least participation in the movement is paying off (*Impact Assessment*) (Table 2.5).

Table 2.5 Resisting the mission and its essential concepts

Resisting the mission: Captures the many reasons for resisting the mission of the initiatives, including commitments to alternative models of youth development, disagreements with leadership philosophies and practices, and the need to see results before committing.

Essential concepts/codes	Definitions
Baggage Assessment	Taking stock of personal issues and biases related to youth and youth development that must be worked through in order to effectively engage with the initiative.
Barriers	Challenges, lack of interest, particular struggles, or other difficulties inherent to the work.
Dissimilarity	The challenges in bringing together different sectors of people, with different perspectives, who normally do not interact, with implications for sharing information, perspective, and resources.
Evidence of Credibility	Research, related evidence, or the opinions of credible leaders are necessary to making the initiative legitimate and credible; the lack of such evidence contributes to resistance.
Incongruity	Discrepancies between participants' ideas regarding the vision and activities of the initiative, with implications for active involvement.
It's Simple But . . .	Although many assets are simple things that people can provide with little effort, day-to-day barriers, including deeply engrained habits, prevent this from happening more regularly.
Joining In	Specific reasons for joining or not joining the initiative, with particular reference to people or sectors that did not immediately jump on the "initiative bandwagon."
Orientation Shift	The shift from deficit- to strength-based thinking and behavior needed to effectively contribute to the initiative.
Owning It First	An individual, organization, or community must take the asset framework, understand it, gain more knowledge about it, and feel passionate about it before promoting it and informing others.

Personal Ownership

This thematic cluster is closely related to the Faith Factor and often comes in the aftermath of Resisting the Mission. In fact, it is arguable that genuine ownership requires a prior resistance if it is to be a tested ownership. We distinguish Personal Ownership from the Faith Factor by virtue of the latter cluster's reference to a deep belief in the Developmental Assets movement and the work of the local initiatives. By contrast, Personal Ownership emphasizes the self—the lived experience of internalizing the asset framework, applying it in one's own life, and living it out in daily interactions. The essence of this code is less on faith in a future investment (healthier youth and a healthier society) and more on an internalized way of being in the world. Personal Ownership references have less to do with faith in the mission and more to do with implications of the framework for one's own life (although, clearly, the two emphases are closely related). One might think of this code as the psychological or secular cousin to the more spiritually imbued Faith Factor. In that vein, the two clusters share a family of essential concepts discussed earlier in this chapter under the Faith Factor, including *Awakening*, *Personal First*, *Orientation Shift*, and *Living It Out*. But within the Personal Ownership cluster, these concepts join with others to capture a sense of self-development, self-understanding, and application to one's own life.

Table 2.6 Personal ownership and its essential concepts

Personal ownership: One must apply the positive shift in attitude toward youth to one's own views and behavior before promoting it to others. Suggests that the larger work of the initiatives can only move forward when participants "live out" the approach in their own lives.

Essential concepts/codes	Definitions
Affirmation	The asset framework reaffirms one's strength-based beliefs and actions, reinforcing one's personal commitment to the work.
Awakening	Recognition of negative biases held toward youth, which, when recognized, results in viewing youth more constructively and strengthening one's personal commitment to the work.
Baggage Assessment	Taking stock of personal issues and biases related to youth and youth development that must be worked through in order to effectively engage with the initiative.
Living It Out	Illustrations of day-to-day actions in one's personal or professional life that depict a clear commitment to the initiative.
Owning It First	An individual, organization, or community must take the asset framework, understand it, gain more knowledge about it, and feel passionate about it before promoting it and informing others.
Personal Gratification	The joy, deep gratification, and personal meaning associated with active involvement in the initiative.
Self-Reference	References to personal or professional experiences that resonate with or introduce one to the initiative, allowing for subsequent deeper involvement.

Personal Gratification is an important concept unique to the Personal Ownership cluster, and, it seems, is important to staying active with the work over the *Long Haul*. This code was frequently linked with *Affirmation*, or the experience of feeling affirmed in one's long-standing orientation toward positive youth development upon learning about the Developmental Assets framework. But personal gratification was not only reported as part of feeling affirmed by the framework, it was also described as an outgrowth of the work. Connecting in prosocial, healthy ways with young people and adults participating in the initiative can be exhilarating and personally rewarding, according to participants from each of the initiatives, particularly when compared to the tiring and often thankless work of treatment or behavior management. Experiencing such benefits at the personal level seems fundamental to owning the framework and incorporating the mission as a way of life (Table 2.6).

Does It Matter?

To paraphrase the chorus of responses related to assessing the impact of the initiatives, "If this is a movement, not a program, how on earth do you evaluate a movement?" Or, as one respondent put it, "How do you measure [this]? Just because [a youth] says, 'Yes, I interacted with an adult,' is that an asset?" While this participant's question takes the complexity of the evaluation challenge to an extreme by suggesting that any interaction between adults and youth might be considered asset building, the point holds. Whereas the ambitions of the movement are grand—changing society's largely negative view of youth and replacing the deficit-based interventions accompanying that view—the work of the movement creeps along in small, consistent increments, according to our study's respondents. *Little Things Count*, they argue, and must be documented if we are to fully appreciate the subtle progress that gets made each and every day. The example of a young person talking with an adult actually would constitute asset building if the conversation were engaging and positive. It can be argued that it is the cumulative effect of these micro, real-world, and nonprogrammatic interactions that make up the core of the mission. But how do we capture these "little things," and the larger cumulative changes they lead to, in a systematic, credible manner?

All eight initiatives encountered formidable challenges to showing that their work had produced a positive impact on the quality of life for youth in the community. The Moorhead Healthy Community Initiative pointed out two assessment dilemmas common to all of the initiatives: (1) the misalignment of traditional funding perspectives with the work of community change and (2) the validity of attributing positive youth and community change to the efforts of the initiative. Time or timing was discussed as the most obvious disparity between traditional funding perspectives and the work of the HC • HY initiatives. Programs are typically funded on a yearly basis, or at most, two to three years, and are expected to produce measurable outcomes, or indicators of change within that time frame. Community change focused on youth development is a long-term endeavor, and the methods appropriate for measuring impact are longitudinal in nature (e.g., changes in community attitudes, infrastructure, and health indicators over the course of a decade or more).

In addition to time, the role of partnerships creates particular challenges for assessment and uncovers further disparities between some traditional funding assumptions and the work of community change. The efforts of the HC • HY communities are collaborative in nature and, when functioning at their best, emphasize the sharing of resources. Funding streams and associated allocation strategies are designed in ways that foster the splintering of community sectors, which, in many cases, must compete for scarce resources. As a result, all of the initiatives have had to work hard to garner to build funder buy-in in order for their sectors to work collaboratively toward the common outcomes they are striving to achieve (Table 2.7).

Table 2.7 Does it matter? and its essential concepts

Does it matter? The various ways of assessing whether the initiatives are having an impact or bringing about meaningful change.

Essential concepts/codes	Definitions
Evidence of Credibility	Research, related evidence, or the opinions of credible leaders are necessary to making the initiative legitimate and credible, and to believing it is having a helpful impact.
Impact Assessment	Specific methods of assessing whether and how the initiative is having an impact on youth and the community.
Little Things Count	Even the smallest efforts when consistently applied make an important contribution to healthy youth development.
Long Haul	An HC • HY initiative is a long-term social and community change effort that could take decades to have a truly meaningful impact.
Not Quite There	Sustaining one's commitment upon recognizing that the initiative is currently falling short of desired goals and expectations.
Survey React	Reactions to the asset survey and related data, which may show that youth are reporting fewer Developmental Assets or community supports than is adequate to promote healthy development.

Community partnerships inevitably lead to questions of "initiative definition," and the attribution of credit for work completed by the initiative's affiliated partners. The Moorhead Healthy Community Initiative provided an interesting example of attribution ambiguity. The initiative helped sponsor a club for young women, supporting it financially and providing valuable networking information for staffing, implementation, and fund-raising. Two years later, the club became independent of the initiative's funding, dissolving any concrete connection between the two entities. Is the good work of this club associated with the initiative? The answers seem to lie somewhere within the definition of the initiative itself. Is it a specific, tangible entity, or is it essentially a movement with a mission that ideally influences the work of other, more tangible entities? The definition of initiative and the assessment of impact are fundamentally interconnected here, and the definition of that interconnectedness, it seems, is open to the perspective of the interpreter.

Finally, it is important to note the relationship between impact assessment more broadly and the specific role of Search Institute's widely used *Profiles of Student Life: Attitudes and Behavior* survey. We created the code *Survey React* to capture the range of responses to the survey, including confusion about its utility as a measure of change over time for Developmental Asset outcomes. Although literature accompanying the survey clearly states that it is not a reliable instrument for measuring community change, the promotion of survey results—particularly positive changes over time—is commonly used as a strategy for mobilizing the community, and as an assessment of impact. As such, anticipated survey results often create expectations that scores will reflect youth who are thriving with a healthy number of Developmental Assets present in their lives. When the results do not reflect this expectation, reactions vary. For some, negative results "are a call to action" and reinforce the view that all youth need focused asset-building opportunities in their lives, not just "those kids over there—the kids engaged in particularly risky behavior," but all kids, including "our kids." For others, negative findings or even small improvements in Years 2 and 3 of an initiative threaten one's faith in the model. Like many funders, they need quick, tangible outcomes.

For some participants, then, the New Norm is fragile—too vague and idealistic. For the believers, it is indeed an ideal, but not a vague one; it is a clearly defined, long-term goal, one only achievable over the long haul. The current status of the eight HC • HY initiatives in this study suggests that the answers to the question of impact assessment are very much individualized, and dependent on what one embraces or is willing to accept as meaningful evidence. In the next section, we describe the dynamic process of working toward the New Norm. How do initiatives act on their convictions, collectively as a movement and locally as individual initiatives, in their efforts to create meaningful progress toward a new way of viewing youth and interacting with them?

Catalytic Context

If the creation of a healthier, community-wide norm for understanding and interacting with youth is the ultimate aim of the initiatives, what are the critical ingredients—resources, processes, and strategies—that have been used to achieve this goal? Based on our synthesis of the broad range of activities and strategies used across the eight initiatives, we created the primary dimension *Catalytic Context* to capture the common core of their work toward community change (see Fig. 2.2). The catalyzing contexts we learned about are environments created by the initiatives through which they strive to build synergistic partnerships among diverse individuals, groups, and community sectors—many of which might traditionally conduct their work independently or compete with one another for scarce resources—in

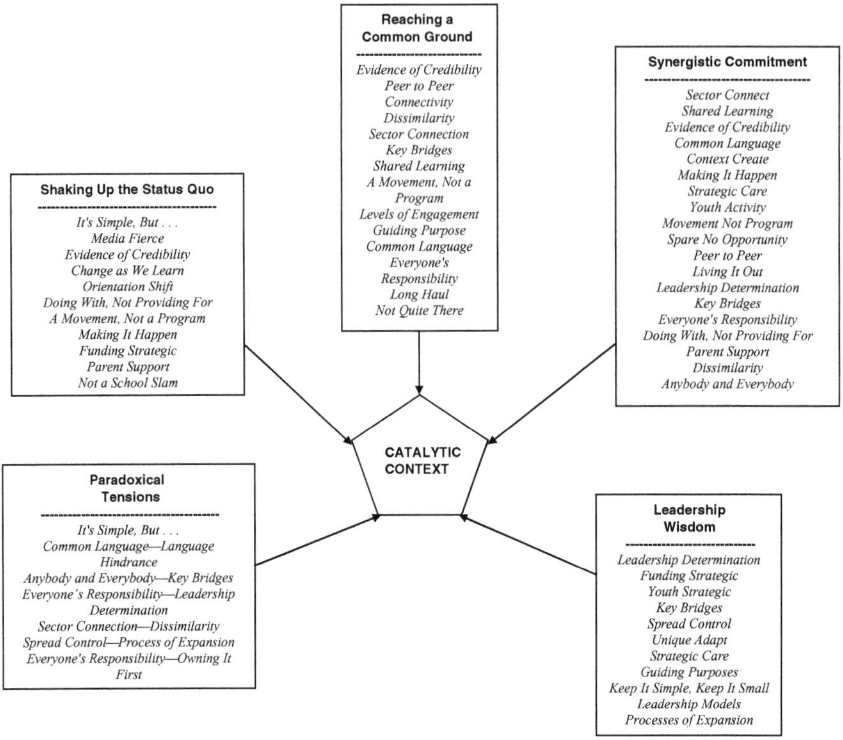

Fig. 2.2 The catalytic context and its thematic clusters

the service of healthy youth development. Through such environments, different constituencies, whether volunteer laypeople, business and community leaders, or professional service providers, find ways to share ideas, expertise, and resources. These dynamic contexts attempt to cultivate reciprocal change among all participants, whereby authentic partnerships across diverse constituencies or sectors produce and reinforce a shift in one's usual way of operating. For example, in the asset approach police officers cannot serve strictly as symbolic representations of discipline and behavior control when partnering with schools; they must actively participate through an asset-based perspective oriented toward constructive relationship building with students, thereby coming to represent safety, support, and encouragement as well as law and order. Schools, in turn, cannot simply invite police officers into their corridors and classrooms to help manage their students; they must actively work with them in a coeducational, proactive manner, creatively tapping into the unique lessons police officers can provide.

Each of the HC • HY initiatives represented in this study fits our definition of Catalytic Context. Respondents from all eight initiatives emphasized the importance of creating a context from which the work of asset-based development could proceed. As will be described in more detail, this context is sometimes more philosophical than physical; that is, the physical space might already have existed, but the philosophical framework for bringing people together around a common mission was lacking. Although this dimension applies to all eight initiatives in our study, the constellation of participating sectors, and the weight of their contributions, varied quite dramatically. As such, the nature of each initiative context is unique, resulting in different manifestations of collaborative activity. In the Moorhead Healthy Community Initiative, for example, the context for the work is centered in out-of-school time with an emphasis on access to constructive recreational activities for low-income and minority youth who historically have been isolated in this community. Key sectors involved include small neighborhood businesses, community policing, and the parks and recreation department, all of which interact to create opportunities that bring recreational resources and advocacy into underserved neighborhoods. The picture is very different in the Traverse Bay Area GivEm 40 initiative, where the work is weighted more heavily in the school settings than in out-of-school-time contexts. Because the school leadership in this initiative made an extraordinary commitment to an asset-building approach, the energy for the work in Traverse Bay has had a predominantly educational flavor. A challenge of GivEm 40 was to broaden the context of the initiative to more fully involve additional sectors.

Catalytic Context is composed of five thematic clusters: *Synergistic Commitment*, *Leadership Wisdom*, *Shaking Up the Status Quo*, *Reaching a Common Ground*, and *Paradoxical Tensions*. Each cluster is characterized by a type of activity, or set of activities, intended to foster the strength-based youth development mission of HC • HY initiatives. These activity sets, when combined, generate the dynamic energy required to envision, pursue, and, ideally, sustain the positive change in community attitudes toward youth (Table 2.8).

Table 2.8. Thematic clusters comprising catalytic context

Thematicclusters	Definitions
Synergistic Commitment	The work of individual people and sectors is exponentially enhanced by a shared commitment to a coherent vision: the whole is greater than the sum of its parts.
Leadership Wisdom	The work of the HC • HY initiatives requires thoughtful, strategic leadership to optimally leverage and utilize contributions from across the community.
Shaking Up the Status Quo	The challenge of breaking the "business as usual" mold for youth development, which tends to be deficit based, short term, and restricted to specialized education and youth development sectors.
Reaching a Common Ground	The challenge of integrating diverse perspectives into a coherent vision that allows all participants to work toward shared goals.
Paradoxical Tensions	Even with a shared vision there are apparent contradictions inherent to the work of positive youth development when divergent sectors collaborate; however, the dynamic tension resulting from these differences can help energize the work.

Synergistic Commitment

This thematic cluster refers to the multiplicative benefits of cross-sector commitment to and collaboration in the work of community change. It implies that the work of individual people and sectors is exponentially enhanced by a shared commitment to a coherent vision. The linking of initiative vision with the "space" for carrying it out is captured through the essential concept *Create Context*, from which this larger cluster originally evolved. *Create Context* explicitly depicts the ways in which participants describe the philosophical core of their initiatives, and the necessity of having such a conceptual center for the work. When describing their HC • HY initiatives, some respondents argued that the essence of the mission is more conceptual than activity based. The activities might look similar to those stemming from an alternative model, but the meaning and implementation of them are fundamentally shaped by the initiative's philosophy. That shared philosophy allows for the tenor and tone of particular activities to be implemented in a manner consistent with the larger goals of the initiative, thereby bringing a sense of coherence to the full range of activities.

Tutoring provides a concrete example of an activity that can be carried out in a wide variety of ways. As an HC • HY activity, tutoring is likely to be seen as a community resource rather than a remedial intervention, and is likely to use a relational approach to learning. *Doing With, Not Providing For* is a concept that signifies the importance of approaching activities with youth in a relational manner rather than constructing the usual "provider–recipient" dichotomies. Even in an activity like tutoring, where there is a clear knowledge differential between the adult and youth participants, many of our respondents would argue that learning *with* youth, even through basic homework requirements, is a fundamentally different learning experience than providing educational support *for* them (Table 2.9).

The synergistic energy behind each initiative's philosophical and working context was generated by a unique constellation of economic, social, geographic, political, and historical factors. The Moorhead Healthy Community initiative was a response to the geographic and social isolation of youth, and the lack of constructive out-of-school-time activities. The Take the Time initiative in Portland grew out of a statewide county government initiative establishing local community commissions to support healthy youth development. One key individual—the local United Way director—who had used the Developmental Assets framework in a prior position spearheaded the GivEm 40 initiative in the Traverse Bay Area of Michigan. The Healthy Community initiative in Orlando grew out of the recognition that high population mobility in central Florida was adversely affecting community building and school readiness for children and youth, who often moved and changed schools several times in the course of one school year. Although these Wave I initiatives had vastly different starting points and progressed along markedly divergent pathways, they communicate and clarify the common need for a coherent philosophical center to provide ongoing direction and conceptual integration.

Although there is a central location, organization, or office for each of the initiatives from which their efforts are organized, the work moves throughout their

Table 2.9 Synergistic commitment and its essential concepts

Synergistic commitment: The work of individual people and sectors is exponentially enhanced by a shared commitment to a coherent vision: The whole is greater than the sum of its parts.

Essential concepts/codes	Definitions
Anybody and Everybody	One does not have to be in a position of authority or a particular status, or possess special abilities to build or share assets.
Context Create	Creating a place, philosophically or literally, to locate and propel the work of the initiative.
Doing With, Not Providing For	Youth are engaged with adults in solving problems or participating in constructive activities, rather than being recipients of services.
Everyone's Responsibility	Positive youth development must be the responsibility of all members of a community, not just a select few.
Making It Happen	The actions carried out within and across sectors in order to realize the specific goals and larger vision of the initiative.
Peer to Peer	An essential aspect of synergistic commitment is realized through youth interaction and outreach on behalf of the initiative.
Spare No Opportunity	Synergy is cultivated through participants' efforts to connect and share information at every feasible opportunity.

respective communities, creating a wide-ranging network of volunteers, service providers, policy makers, neighborhood associations, school committees, and city councils. *Making It Happen*, the code summarizing the enormous range of activities that move the initiatives forward, is, according to our respondents, the work of *Anybody and Everybody*. And, as discussed earlier, rather than deferring the work of healthy youth development exclusively to parents, teachers, and trained professionals, it must be taken on as *Everyone's Responsibility* if community-wide transformation is to occur. But for anybody and everybody to take on his or her share of responsibility for making change happen, the presence of the initiatives and the diversity of activities they provide must be contextualized within a coherent vision. The creation of a space for that vision, philosophically and literally, allows for the commitment of energy invested in all corners of the community to have a synergistic impact rather than one diffused through disconnected efforts.

Orlando's Healthy Community Initiative presented the challenging balance it must maintain in promoting and supporting neighborhood ownership of projects, while simultaneously keeping the neighborhoods and initiatives connected to a common vision and philosophy that underlie the range of projects. As its director puts it, there's an important tension between neighborhood ownership and initiative cohesion:

> We have to work really hard and be very conscious of trying to keep a cohesiveness to what's going on here so that there's some rhyme or reason to what's going on in this part of town, to that part of town, so that people know about it and each other . . . The truth is if a meteor comes and hits my organization, Winter Park and West Orange will continue with their initiative. So it's in effect self-sustaining and very locally owned. At the same time, we work hard to be kind of the center that holds . . . and so we have a community asset network.

We have a Web site. We facilitate. We connect. We do all the things so that overall there is a strategy and, you know, a cohesive strategy for our community as a whole, but the way that we're going about it is by nurturing these kind of subcommunities in building assets.

Leadership Wisdom

The words of Orlando's director lead naturally into our next thematic cluster, which emphasizes the strategic decisions made by initiative leaders. We highlighted "wisdom" in this leadership description, because the decisions made tended to be the product of years of experience in community and organizational change efforts. Wise leadership came across as vital to the progress and sustainability of each initiative, as would be expected. It seemed particularly helpful for initiative leaders to have learned from past mistakes in other arenas, and to use that experience to prevent the repetition of similar mistakes in the current work. The concept *Strategic Care* captures the intentional actions of initiative leaders. It is not enough to care deeply; one must invest in caring activities planfully and collectively in order for the good intentions to be fully realized (Table 2.10).

Several other essential concepts link with *Strategic Care* in the exercise of wise leadership. In combination, the two concepts *Processes of Expansion* and *Spread*

Table 2.10 Leadership wisdom and its essential concepts

Leadership wisdom: The work of the HC • HY initiatives requires thoughtful, strategic leadership to optimally leverage and utilize contributions from across the community.

Essential concepts/codes	Definitions
Funding Strategic	Strategies for raising financial support and for the use of funds, as well as strategies used by funders to support the initiatives.
Guiding Purposes	Core philosophies, beliefs, and values that guide the initiatives.
Keep It Simple, Keep It Small	Emphasizing simplicity in recognition that people can "burn out," give up, or lose confidence when initiatives seem too complex.
Key Bridges	Critical connections across sectors facilitated by key individuals, organizations, or groups.
Leadership Determination	A strong champion, organization, or group that exercises the experience and wisdom necessary to moving the initiative forward, often in the face of extraordinary challenges.
Leadership Models	Intentional approaches to managing or guiding the initiatives.
Processes of Expansion	Ways in which the initiative develops through intentional strategies or unintentional events or circumstances.
Spread Control	The need for choosing certain directions over others, and making choices that eliminate some directions while prioritizing others.
Strategic Care	Strategic approaches to planful action; exercising judgment or discretion in decisions perceived as having a powerful impact.
Unique Adapt	Adapting the framework to the uniqueness of different communities, sectors, or groups of people.
Youth Strategic	Thoughtful actions and ideas carried out or proposed by youth.

Control constitute a critical area within which leaders must make strategic decisions. Given that the initiatives are intended to be community-wide, there is a common impulse to expand the work quickly and involve as many sectors as possible. The risk with this approach, according to one of the leaders in the Traverse Bay Area initiative, is the creation of an impact that is "an inch deep and a mile wide." Optimal, sustainable community impact, from her perspective, can best be fostered through careful within-sector growth, allowing lessons to run deep within a sector prior to sharing and expanding upon them across sectors. That view, however, is not held by all initiative leaders. Some endorse a more organic model of growth stemming from carefully planned exposure to the Developmental Assets framework for as many constituents as possible. The diversity of experiences among these constituents, they argue, will lead to approaches to building assets that could not be envisioned or generated by a more narrowly defined group of participants. Whatever the strategic plan for community involvement, the tension between expansion and being spread too thin is confronted by all of the initiative leaders.

Adapting the ideas associated with the larger HC • HY movement to the needs of the local context (*Unique Adapt*) is another leadership challenge common to each initiative. In some cases, the model's local fit can be a particularly hard sell. The concept *Leadership Determination* depicts the extraordinary efforts of leaders to address those challenges and meld the asset framework into the local consciousness. The challenges to adapting and applying the framework are many, including resistance from within the local service-providing community. "Turf protection" can be especially difficult to penetrate when there is heated competition among programs for limited funding. In such cases, *Leadership Determination* may be manifested in negotiating partnerships among programs that would not typically collaborate. It also is expressed through forging relationships with local funders (*Funding Strategic*) aimed at persuading them to distribute resources in a manner consistent with the tenets of the asset framework, including the prioritization of partnerships among divergent sectors. In a host of different ways, determined leadership is present in all eight initiatives; it is that component of Leadership Wisdom that recognizes the need to bear down and pursue selected goals through the sheer will of one's convictions.

Shaking Up the Status Quo

Inherent to catalyzing change, according to the initiative leaders, is a shaking up of the community status quo. Determined HC • HY leaders indeed tend to be key status quo shakers. Building the HC • HY initiatives involves confronting entrenched ways of thinking about youth and the systems built upon that thinking. The concept *Orientation Shift* encompasses leaders' recognition of the need for dramatically altering both thinking and action at all levels of the community. Taking this recognition beyond thought to action, and sustaining that focus over time can be a demanding and relentless task. It includes shifting the perspectives of funders (*Orientation Shift—Funders*, and *Funding Strategic*), as described earlier. Funders not only must be convinced to support the healthy development of all youth rather than simply providing resources for those most "at risk," they must also buy into supporting *A Movement, Not a Program*. The latter commitment constitutes a substantial leap of faith, particularly given that they are encouraged to commit support for an extensive period of time rather than for shorter, more discrete time periods, as is the norm for program funding (Table 2.11).

Table 2.11 Shaking up the status quo

Shaking up the status quo: The challenge of breaking the "business as usual" mold for youth development, which tends to be deficit based, short term, and restricted to specialized education and youth development sectors.

Essential concepts/codes	Definitions
Change as We Learn	Initiatives learn from their successes, challenges, and mistakes, resulting in shifts in direction or action.
Doing With,	Youth are engaged with adults in solving problems or *Not Providing For* participating in constructive activities, rather than being recipients of services.
Everyone's Responsibility	Positive youth development must be the responsibility of all members of a community, not just the usual "specialists."
Evidence of Credibility	Making change requires research or related evidence in support of the legitimacy and credibility of the initiative's efforts.
Funding Strategic	Strategies for raising funds must shift from short-term, "objective" relationships with funders to longer-term partnerships.
Impact Assessment	Similar to Evidence of Credibility, making and sustaining change requires targeted assessment of specific initiative actions.
It's Simple, But ...	Although many assets are simple things that people can provide with little effort, day-to-day barriers, including deeply engrained habits, prevent this from happening more regularly.
Media Fierce	Altering the negative portrayal of youth in the media.
A Movement, Not a Program	Shifting youth development thinking away from short-term programmatic interventions to ongoing community commitments.
Orientation Shift	The shift from deficit- to strength-based thinking and behavior at the individual, institutional, and community levels.
Not a School Slam	Moving away from holding schools excessively accountable for the comprehensive development of youth.

A challenge previously addressed in the Resisting the Mission cluster, which also is central to Shaking Up the Status Quo, hinges on the nature of outcomes and evidence used to document the initiatives' success (*Impact Assess* and *Evidence of Credibility*). Funders seeking short-term outcomes tied to discrete program interventions must have their long-held commitments to interventions and assumptions about service efficacy challenged if they are to participate meaningfully in the life of the initiatives. One specific challenge to funders, and the communities more broadly, is that of shifting excessive responsibility for poor youth development outcomes away from the schools. *Not a School Slam* is a concept that emerged out of the work in Orlando, where strong efforts are made to distribute the initiative's focus across a wide range of sectors. The schools, according to the initiative director, are typically charged with the undue burden of addressing the whole array of needs and issues students bring with them, and are criticized excessively when the outcomes for highly stressed youth are predictably bad. His stance set an important tone for our understanding of HC • HY work in Orlando and beyond, with the emphasis on holding the whole community responsible (*Everyone's Responsibility*) for healthy youth development, rather than blaming schools—or any other single institution—for their deficits.

The status quo shift that perhaps has been the most difficult to realize is the move from providing services for youth to working actively with them (*Doing With, Not Providing For*). Youth services, when clearly planned, are specific and straightforward. But what does it really mean to join with youth in the work of community change? Portland provided one of the most compelling examples through its strategy for altering the media's portrayal of young people. Rather than relying on newspaper reporters to submit inspiring stories of youth activity, the young people began their own column in the local newspaper. It included stories written by youth about youth, and required collaboration between the young people and the newspaper staff. Despite such examples, however, the initiatives tended to struggle in their efforts to broadly implement approaches to adult–youth collaboration. At the heart of this struggle, according to participants, is society's impulse to dramatize crisis and fixate on bad news in general. Consistent engagement in healthy behavior tends not to be newsworthy in our society—"If it bleeds, it leads," as one participant put it. Confronting this aspect of the status quo has proved to be among the more relentless tasks of the initiatives and their leadership, including their youth leaders.

Reaching a Common Ground

Healthy working relationships among disparate community sectors are essential to creating the catalytic activity needed to foster a new norm for perceiving and interacting with youth. Two concepts—*Sector Connection* and *Dissimilarity*—were among the most frequently cited as central to reaching a common ground for this work. The first is a more general concept, referring to the myriad ways in which sectors connect with the initiative and with each other. Connection with the initiative helps cultivate the coherent community-wide vision cited under Synergistic Commitment. The second code, *Dissimilarity*, is a specific form of sector connectivity; it refers to breakthroughs in planning and action that occur when sectors with markedly different core functions come together and collaborate on behalf of healthy youth development and community change. It also captures struggles that ensue when divergent sectors attempt to work collaboratively.

Reaching a Common Ground is distinguished from Synergistic Commitment by its emphasis on the challenges inherent in getting to a common place from which one can make a commitment to the mission. Synergy implies positively energized working relationships, which summarizes what the HC • HY initiatives are working toward, and what they experience when functioning at their best. But getting to that place of peak synergistic performance is hard work, and often includes intense and painful struggles. In addition to hearing inspiring stories of collaboration, we heard many stories of turf warfare and resistance to the work of the HC • HY initiatives. Whereas Synergistic Commitment depicts the exciting plans and activities jointly constructed by collaborating sectors, Reaching a Common Ground captures the seemingly mundane, everyday persistence required in moving the work forward. Many of the essential concepts converge around this dimension in a particular way. For example, *Making It Happen* contributes snapshots of the little things necessary to move the work forward, such as endless meetings, plans for reaching out to new pockets of youth and new sectors, and painstakingly seeking ways to collect *Evidence of Credibility* to reassure doubters that the new direction is worthwhile and that progress is being made (Table 2.12).

Table 2.12 Reaching a common ground and its essential concepts

Reaching a common ground: The challenge of integrating a wide range of perspectives and approaches into a coherent vision that allows all participants to work toward shared goals.

Essential concepts/codes	Definitions
Common Language	The assets language facilitates shared understanding.
Connectivity	Ways in which individuals connect around a shared vision.
Dissimilarity	The importance of a shared vision for integrating the work of diverse people, sectors, and points of view.
Everyone's Responsibility	Reaching common ground requires the commitment of all.
Evidence of Credibility	Research and related evidence are necessary to shaping a shared vision and retaining faith in it.
Guiding Purpose	Initiatives are guided by a commonly understood focal purpose.
Key Bridges	Influential people or groups who rally diverse sectors of the community around a common goal.
Levels of Engagement	The degrees to which people are committed to the vision.
Long Haul	Reaching common ground requires long-term commitment.
Making It Happen	Actions taken to move the initiative toward a common ground.
A Movement, Not a Program	Organizing around a common vision rather than distinct programs.
Not Quite There	Acknowledgment of not yet realizing the shared vision.
Peer to Peer	Youth working together to achieve the initiative's goals.
Sector Connection	Ways in which sectors connect around the shared vision.
Shared Learning	Exchanging strategies for achieving common goals.

Common Language and *Long Haul* are two additional concepts that contribute uniquely to this thematic cluster. Getting to a place from which the work can be synergistically conducted requires a language that allows for mutual understanding. For the message to be carried forward on a daily basis, that language must be engrained in the community's consciousness, competing for attention and reflective space with the dominant societal images of youth that are so pervasive. And, finally, while Synergistic Commitment represents a powerful ideal worth pursuing, getting to that point may take awhile, even years. The concept *Long Haul* represents the belief held by some participants that the struggle to reach common ground may be endless or ongoing, and that this struggle will necessarily coexist with the exhilaration experienced through synergistic accomplishments. In short, struggle and exhilaration are part of a singular process, from this perspective, creating a productive tension that is part and parcel of the work.

Paradoxical Tensions

This final cluster organizes those codes that on the surface appear unrelated or even contradictory, but which, upon closer examination, influence the very definition of each other through their interaction. The most pronounced paradox that emerged from our study is represented through the concept *It's Simple, But* .... This concept is intended to depict the paradoxical reality of the Developmental Assets framework being simple and straightforward, a conceptual approach designed to be comprehended and carried out by *Anybody and Everybody*, and, at the same time, often being difficult to act on consistently due to barriers in time and energy, and the entrenchment of a status quo view of youth that in many respects sits in opposition to the core messages of the framework.

Responses from all eight HC • HY initiatives point to the dialectical tension between *Anybody and Everybody* and *Everyone's Responsibility*, on one hand, and *Leadership Determination* and *Key Bridges*, on the other hand. Effective and persistent leadership is essential to supporting and sustaining the everyday work on the ground. Additionally, the ability to leverage the power of well-situated key individuals—such as school system superintendents, political figures, and business leaders—was reported as pivotal to opening up new possibilities for the initiatives. The juxtaposition of the simplicity of the asset framework with the need for powerful leadership and access to community resources captures the nature of the initiatives much more clearly than either part of that tension alone (Table 2.13).

The dialectical relationship between the expansion and manageability (*Processes of Expansion* versus *Spread Control*) of the work also was a prominent theme among study participants. A fundamental tenet of the Developmental Assets framework is that efforts to build a healthy community on behalf of healthy youth should be a community-wide undertaking. But moving too quickly with the HC • HY work can lead to burnout among initiative participants and a superficial application of the work. The relationship between breadth and depth, therefore, has important implications for the timing of plans to target new sector involvement, outreach to youth, and funding priorities. Ideally, each initiative ultimately would be both broad and deep, but the selection of pathways toward that goal vary extensively across initiatives.

Another paradox described previously is the relationship between *Common Language* and *Language Hindrance*. The Developmental Assets language serves to provide both a common vocabulary for understanding youth development and the work needed to promote it, while at the same time it can unintentionally create roadblocks to understanding. Not everyone translates "asset" language into human development terms as readily as might be expected, and, as explained in our description of the New Norm, some are opposed to the "capitalist" implications in phrases such as "investing in youth as community assets."

Table 2.13 Paradoxical tensions and its essential concepts

Paradoxical tensions: Even with a shared vision, there are apparent contradictions inherent to the work of positive youth development when divergent sectors collaborate; however, the dynamic tension resulting from these differences can help energize the work

Essential concepts/codes	Definitions
Common Language—Language Hindrance	The assets serve as a common language while also proving cumbersome at times.
It's Simple, But . . .	Although many of the assets can be provided easily, consistent provision is difficult by virtue of engrained habits.
Anybody and Everybody—Key Bridges	Although everyone can participate effectively, particularly influential people are critical as well.
Everyone's Responsibility—Leadership Determination	While everyone has a role in the initiative, highly committed leadership is required to keep things moving forward.
Sector Connection—Dissimilarity	Although there is a common vision regarding which sectors connect, their diverse professions make this highly challenging.
Processes of Expansion—Spread Control	Just as initiatives need to reach out across the community, they must also prevent the work from spinning out of control.
Everyone's Responsibility—Owning It First	Although everyone is expected to take responsibility for the initiative, it's critical that they apply it in their own lives first.
Little Things Count—Impact Assess	While every action is important to the larger initiative, careful assessment must target specified outcomes.

Conclusion and Implications

In carrying out our analysis of the eight HC • HY initiatives, we encountered an extremely complex and often paradoxical set of youth development approaches, processes, and systems. The interpretive findings presented in this chapter represent our attempt to capture in clear and useful terms the ways in which initiative participants described *how* they work together, *why* (toward what ends and for what purpose or mission), and with *what outcomes* or goals in mind. In particular, we sought to capture what is uniquely transformative about the Developmental Assets work of the HC • HY initiatives—for the individual participants, their agencies or organizations, and their communities as a whole.

In conducting the analysis, we also strived to create an analytic model that allowed us to address cross-initiative characteristics and processes as well as those qualities unique to each initiative. We see the utility of the model derived from this research as rooted in its ability to simultaneously capture the general and the specific—to help us better understand the world of youth and community development work that these initiatives epitomize, as well as the uniqueness of singular initiatives that can only be captured through case studies. A risk in presenting a general analytic model, even one developed from the ground up as is the case here, is that the model may mistakenly be applied to other initiative work as a road map or guide. We prefer to view the outcome of our analysis as a working model of the widely varied applications of the Developmental Assets framework in different HC • HY communities, one most appropriate for examination and critique at this point. Our hope is that the analytic model will inform other initiatives rather than guide them. Because the findings are derived from the work of experienced HC • HY initiatives, they include elements that may prove helpful in anticipating future opportunities, decision points, and roadblocks.

The richness of the data shared by our study participants could not be adequately or accurately synthesized into a simpler set of findings, although we tried. The process of building an analytic conceptual model that accurately represents the quality and quantity of information shared has taught us yet again the value of embedding specific findings within a salient context. To return to our initial research questions, our knowledge of how—and even if—these HC • HY initiatives shape youth development will be comprehensible only if embedded within a fuller picture of the processes by which the initiatives are formed and maintained, and through which individual and collective transformation is achieved.

The next step for this research project is to move from description of processes to assessment of results. How does this conceptual model of our findings help us not only to describe the initiatives but also to assess their ultimate effectiveness? Can the model be used as a tool for youth and community development evaluation that links process and outcomes, descriptive context and "hard findings"? If so, how? To help answer these questions, our next step is to apply the conceptual model within particular initiatives. Whereas the process presented here was largely inductive (building a model from the ground up), our next steps will be deductive (applying the model to existing initiatives to assess its utility in learning not only what initiatives have done to foster positive youth development but also to assess how effective those efforts have been).

In the community case study chapters that follow, we show how each of the eight HC • HY initiatives uniquely contributed to the findings from our study. While this chapter has laid out the thematic structure of our findings, the case study chapters bring the findings to life through the many voices of the youth and adults involved in the actual activities. Our hope is that by seeing the analytic model evolve and demonstrated through the efforts of communities across the country, the dialogue necessary for expanding this critical work will be enhanced. In other words, like the HC • HY initiatives themselves, we hope our depiction of their efforts will serve as a catalyst—not for replicating what is already being done, but for advancing the innovation and effectiveness required to transform collective actions on behalf of youth development.

Chapter 3
Strategic Care, Sector by Sector: Traverse Bay Area's GivEm40 24.7

We begin our presentation of the eight Healthy Communities • Healthy Youth (HC • HY) initiatives with Traverse Bay Area's GivEm40 24.7 (GivEm40) to illustrate an initiative that intentionally chose to go sector deep versus spreading its efforts equally across a multitude of community sectors. The hallmark of GivEm40 is the exceptional buy-in of the educational system, with all 19 of the area school system superintendents making strong commitments to the initiative. By beginning our presentation of the HC • HY approaches with a model that emphasizes sector-specific depth as its starting point, we are positioned to build toward applications that prioritize more complex cross-sector collaborations.

Context of the Initiative

The GivEm40 coalition in northwestern Michigan formally launched its initiative in 1999, bringing together a group of key stakeholders across a five-county region. The founding leadership group labeled the initiative a *coalition* to underscore the distinction from programs or individual youth-service sectors and to emphasize the need for organized collective action. Since the inception of GivEm40, youth and youth-serving organizations, faith communities, media, law enforcement, family-serving organizations, school districts, and businesses have worked together to heighten receptivity to and awareness of positive youth development in Traverse City and the surrounding Traverse Bay Area. The early and extensive efforts to mobilize support for the work of the coalition were rooted in prior experiences of promoting positive youth development, which taught GivEm40's leadership team the necessity of building a critical mass of momentum. As the coalition evolved, however, it emphasized building momentum one sector at a time rather than rallying support across the entire community simultaneously. In this sense, the GivEm40 story is one of going deep and gradually spreading out from that anchored starting point versus going broad and deepening the efforts wherever possible.

Northwestern Michigan has a population of approximately 145,000 residents and consists of five counties: Kalkaska, Benzie, Antrim, Leelanau, and Grand Traverse. Located at the northern tip of Michigan's Lower Peninsula, the area is roughly

M.J. Nakkula et al., *Building Healthy Communities for Positive Youth Development*, The Search Institute Series on Developmentally Attentive Community and Society 7, DOI 10.1007/978-1-4419-5744-3_3, © Springer Science+Business Media, LLC 2010

the size of the state of Delaware. Most of the population (75,000) lives in Grand Traverse County, and close to Traverse City, the hub of the region. Similar to Orlando (see Chapter 6), but on a much smaller scale, Traverse City is a resort town, dependent on a service economy. The beaches, protected parklands, and ski slopes draw an affluent population. However, the number of students receiving free or reduced-cost lunch in the five-county intermediate school district exceeds the state average, due to the lower-income rural communities that surround Traverse City. Similar to Orlando, the region had undergone a period of rapid growth for an extended period prior to the onset of its HC • HY initiative, which helped signal the need for a more coherent approach to supporting the area's youth.

Structural Features of the Initiative

GivEm40 is administered through the United Way of Northwest Michigan, which, at the time of our study, was led by executive director Becky Beauchamp, who is credited with bringing the Developmental Assets framework to the area from her previous work in Michigan's Upper Peninsula. Prior to the launch of the initiative in 1999, three major funders had committed financial resources and staff time to help get the initiative off to a strong start, and all 19 of the region's public school districts agreed to administer Search Institute's *Profiles of Student Life: Attitudes and Behaviors* survey to students in grades 7, 9, and 11. Beauchamp insisted on a "critical mass" of commitment from the region's school administrations, funding sources, media, and major university prior to the launch of GivEm40, and the coalition was deliberate in targeting the leadership from these constituencies.

GivEm40 is unique among the eight HC • HY initiatives in our study in its partnering with a university research and evaluation group—Michigan State University (MSU)—from its inception. Several MSU faculty members who specialize in evaluation are members of the coalition and have developed a logic model for assessing the work. Having a university-based research team as part of the initiative rather than as an external evaluator provides in-house guidance for thinking about evidence-based approaches to the work, including what might be considered persuasive results. As discussed in Chapter 2, *Evidence of Credibility* is a key concept/challenge that runs through all the HC • HY initiatives we explored (the thematic clusters and concept names are presented and defined throughout Chapter 2). The MSU evaluation group has helped GivEm40 address that issue.

As noted earlier, GivEm40 approaches the dissemination and implementation of the Developmental Assets model in a sector-specific way, with the goal of achieving depth rather than breadth. A clear statement of a sector's deep commitment to the initiative was the hiring of an asset development coordinator by the Traverse Bay Area Intermediate School District. By creating and funding such a role, the district sent a strong signal that its leadership was unified in its support of a Developmental Assets approach, and it created a structural reality that made the school district the early center of the coalition's work.

Characterizing Themes

As an initiative that emphasized sector-specific depth, there are particular qualitative themes from our study that were prominently displayed in GivEm40. It is noteworthy that this sector-specific approach did not occur by happenstance. Rather, Becky Beauchamp and her colleagues thoughtfully planned for the sector-by-sector method reflected in the thematic summary that follows.

Leadership Wisdom and *Representation* are the two thematic code clusters featured in our portrayal of GivEm40. The coalition has vivid lessons to teach with respect to integrating the core of these two clusters. As we read the portrayal that follows we might ask, for example: Who is represented in the leadership of the coalition? Whom is the leadership representing? How are decisions of inclusion and exclusion made? Such questions are important in understanding leadership challenges for all HC • HY communities.

Leadership Wisdom

As shown in Table 10 of Chapter 2, Leadership Wisdom suggests that the work of the HC • HY initiatives requires thoughtful, strategic leadership to optimally leverage and use contributions from across the community. The GivEm40 coalition is replete with examples of Leadership Wisdom, as reflected in the following selection of essential concepts that contributed to this thematic cluster for the Traverse Bay Area.

Strategic Care

This essential aspect of Leadership Wisdom captures the importance of working smart and strategically. It assumes that for care to be provided most usefully, it must be accompanied by thoughtful planning. The concept is an antidote to spontaneous "do-gooding," which the GivEm40 coalition views as "feel-good" caring at the cost of meaningful, sustaining care and support. The label of the code intentionally blends "strategic planning" with care and support, to emphasize the critical importance of strategy in the caring work of the HC • HY initiative.

GivEm40 was instrumental to our understanding of leadership as part of *Strategic Care* and in Developmental Asset work more generally. This coalition explicitly focused on engaging leaders who were critical to the progress of the initiative, prior to its community-wide launch. As one member put it in a community meeting we attended, "You can't just go after the low-hanging fruit, but need to go after the blossoms—the power brokers." Establishing links among sectors, through coordination of sector leadership, was seen as the critical mobilizing effort that would differentiate the Developmental Assets movement from previous programs that had experienced a "shelf life":

> People say, "We've been doing this forever, why do we need a coalition?" Ah, and it's precisely about the community conversation of, if we stay with a bunch of isolated, disparate efforts in the community, obviously the sum total of that is not producing the results we want

for children. [T]hat is a harder conversation to have in a regional community than I ever anticipated . . . but, you know, we're getting there. It's about, you know, if the schools work in their beleaguered silos, and if community organizations work like the world revolves around them, and if we don't have citizens engaged in this process, and if we don't have more of a community road map, we're just going to get kind of the same result. And . . . somehow this is . . . a conversation about the sum total being more than just the individual parts.

A key component of Strategic Care is helping different community sectors see the uniqueness and importance of their role within the larger picture of the initiative's work. Our code *Sector Connection* refers to the specific ways in which sectors from the eight communities make their contributions. Traditionally, youth service agencies have represented an inordinate proportion of the community's efforts to support the development of its young people. GivEm40, like all the eight initiatives in our study, has strategically worked to bring more balance to this picture. A participant in a coalition staff meeting commented the following:

[We're] trying to have the agencies see that they're a piece. They're an important piece where they link, but they're not the whole puzzle or half the puzzle. And while that seems like such an easy kind of conversation, it is not . . . because everyone is used to doing their own thing.

Coordination of sector-specific leadership in GivEm40 was often facilitated by positive, long-term relationships among colleagues. The regional school district superintendents' ability to work together and gain the support of all their schools is such an example. This educational leadership group had convened over time to discuss potential strategies for more coherently supporting students' development. Careful attention to timing, strategy, and action characterizes the larger coalition's leadership, and there is a sense that little occurs accidentally. There are many descriptions of the importance of judgment and timing, such as introducing the area's "pilot schools initiative," which served to stimulate the larger district's involvement.

We sold these pilot schools with Youth Friends [a mentoring program], and we bridge-granted them and we painted all these rosy scenarios . . . We didn't have a dime of it raised . . . ah, but that's how we sold the pilot schools. Now we can separate them [end the bridge grant] because the school's seeing, "Wow, there's some things that can happen." We got us some superintendents ready to talk about what can happen in schools. We don't have to [incentivize] them anymore to do it. We don't have to oversell on Youth Friends [or any other specific aspect of the initiative] . . . We couldn't have done that a year ago. (Focus group participant)

"Strategic abandonment" is a term the coalition coined that captures the necessity of choosing some efforts over others to counter the tendency, described by all of the initiatives, to overextend themselves. GivEm40 was a relatively young initiative compared to the others in our study. Yet, in its first two years it had grasped the pitfalls of expansion, which may be an important experiential criterion for effective and sustainable initiative leadership. It was stated in a staff meeting: "We had to get real about the amount of time that it actually takes to do one thing well, and go deep versus going a mile wide . . . guess it's that strategic abandonment thing."

GivEm40 alerted us to the common HC • HY experience of feeling "spread too thin" by virtue of the need to reach all sectors and address all 40 assets. Although the long-term goal may be to reach all sectors and address the assets comprehensively, the process of getting there requires careful planning and takes time:

> "Good idea. It sounds like fun, sounds cute, sounds like it could work. Let's go there." And you'll be spread all over the map, ah, and not have a strategic focus, and not gain that momentum that's going to go, that's really going to set down roots. (Focus group participant)

Going big by starting simple is another way of describing GivEm40's strategic focus. *Keep It Simple, Keep It Small* is the code we developed to capture a tenet that emerged through our study, namely, that application of the Developmental Assets framework must remain simple enough to grasp by everyone potentially engaged in the work. If it becomes too complex and prematurely expansive, it will not be manageable and may alienate people rather than engage them. If it stays simple enough to grasp and small enough to recognize, then an increasing number of people will get on board. Paradoxically, then, growth will occur through the spreading of simple, local efforts that ultimately make up the larger community quilt of Developmental Assets. Keeping it simple and small is not an effort to reduce growth but rather to nurture the expansion of an initiative wisely so that it can reach the entire community in a natural, manageable way. As a focus group participant put it, "[We will grow best by] focusing on a few things, doing them well and learning from them, and figuring out how to expand those rings of concentric circles."

Leadership Determination: The Element of Risk Taking

An important part of leadership is determining when to be cautious and when to take risks. Each initiative made unique contributions to our understanding of effective leadership for community youth development. Some emphasized the importance of local, grassroots leadership and discussed processes for developing it. Others pointed to the extent to which leadership represents the constituencies not traditionally represented. GivEm40 taught us particular lessons about strategic risk taking. Many of those lessons are embedded within the coalition's particular application of *Leadership Determination*, which underscores the need for initiative leaders to forge ahead in the face of resistance, apathy, or lack of understanding. In general within our study, *Leadership Determination* joins with *Strategic Care* in comprising the heart of the Leadership Wisdom cluster. Once strategy has been articulated, determined leaders find ways to accomplish the initiative's agendas, despite the many barriers encountered along the way.

GivEm40 described strategic risk taking as a "high-wire act." Leaders must be willing to take risks, but they also need to have an articulated plan and be cognizant of possible consequences of risky decision making. As the following quotation from the initiative's director depicts, however, lack of risk taking may carry its own consequences:

I think the assets are [in] danger of being a nice blather in communities, a very nice blather. Or, you know, is it really going to change the environment for children? We're on the verge of this really becoming something here. Or on the verge of it just having been, "Oh, this nice thing" that we can talk about and congratulate ourselves about, but that really didn't accomplish ... all that it can. And that's the part that I think is really tenuous ... I have no sense of illusion about what could happen to this. I mean, there's ... you know, three ... bad decisions in the next week, you know? I mean, just things I see happening in the next week, could, ah, you know change the course of [things ... with] ... this really becoming, you know, an Amtrak train moving through the community.

Through initiatives like HC • HY, in which leadership typically is distributed widely owing to the multisector nature of the work, the risks too must be shared. Who better to address this than a program officer of a partnering foundation?

Something I've learned ... in this process of helping to form a coalition ... is that partnership really is the work of, ah, secure individuals from mature organizations that have the ability to take a little bit of risk.

Within GivEm40 perhaps the largest risks were taken by the leadership in the educational community. Beauchamp describes the risky decision of re-administering the *Profiles of Student Life* survey two years after it was first administered, knowing there were slim odds of obtaining positive results in such a short time. But not to have administered the survey again would have slowed momentum, and potentially defused the school systems' emphasis on the assets as an organizing framework for growth and development. So they took the risk and dug in to argue the long-term benefits in the face of potential short-term disappointing news:

But that kind of skepticism is real, it's what you face in communities ... you can sound like you're backpedaling ... And that's just the high-wire act without a net. Even with the schools ... I mean the ISD [Intermediate School District] is a wonderful partner in this thing, but it takes so much risk on their part to say, "Okay, we're going to commit these resources." It's pretty hard for an organization to keep its attention on something with no money attached to it, you know, with more dreams than actual substance at the moment. To stay focused here, to keep going in that work. I mean that takes ... you know, an enormous amount of faith, ah, that most organizations don't have. (Staff meeting participant)

Leadership Resistance

Maturity is a prerequisite for effective initiative leadership, but it can also be a barrier to initiative progress if key leaders are reluctant to change in order to maintain influential positions in the community. Mature leaders typically have worked in their positions or fields for many years, and often have worked from the deficit orientation that has been characteristic of youth work. An aggressive emphasis on what is not working runs the risk of devaluing those who have been working in, leading, and funding youth programming for many years:

[I]t does threaten some of them. "What we've done to date doesn't count." [A]nd there's just the continual discussion, communications, [which we must engage in] without getting so totally frustrated, because it feels like they're dragging us back. And from the community

> perspective, the people I anticipated being leaders have clearly been some of the ones that are dragging their heels. And that's disappointing. It's probably predictable ... [this need to] carve out their turf. (Focus group participant)

Leadership resistance is related to our thematic cluster Resisting the Mission, which is emphasized more fully in Chapter 10. This cluster depicts various reasons for resisting the Developmental Assets approach, including doubt rooted in a lack of systematic evaluation research (due in part to the unique nature of each HC • HY initiative) and lack of knowledge about the framework. But clearly a key component of such resistance, in many cases, is dogmatic and entrenched leadership. As members of GivEm40's leadership team noted, even those sector or community leaders who might have taken strong action on behalf of youth historically are capable of resisting partnership in the asset-based work: "you're even seeing the people who are bold partners ... almost derail it, because it's threatening." The threat, according to HC • HY leaders in Traverse Bay and across all eight initiatives, is based partially in fear that joining the initiative means letting go of past practices and the predictable results associated with those practices, even if those results are less robust than ideal.

A fundamental aspect of Leadership Wisdom, and a key means by which resistance is addressed, is represented in our code *Unique Adapt*, which depicts the importance of adapting the Developmental Assets framework to the needs and structural realities of each community. By respectfully introducing the framework to strategically targeted community leaders, and adapting it to specific community needs and existing youth development and support systems, certain aspects of resistance can be defused. In each initiative, certain sectors were easier to engage because they perceived themselves to be in philosophical agreement and behavioral alignment with a positive youth development orientation. In other cases, initiative leaders take great pains to present the Developmental Assets framework in a manner that will be congruent with the larger goals and practices of a particular community sector.

GivEm40 was particularly sensitive to aligning the asset framework with existing practices in various community sectors, and as a result was quite successful in mobilizing several existing systems to reorganize according to a Developmental Assets orientation. Foundation officers, university faculty, law enforcement officers, and school superintendents all described themselves as participants in systems that originally were not aligned with the framework. A program officer representing Rotary Charities described the atypical nature of the foundation's partnership with GivEm 40:

> The concept of us as a foundation being a part, a working part, of a coalition was very different than what we've done in the past. [I]t has [been] a different perspective, and it's had some challenges from our board's perspective, in that my staff time usually isn't allocated to participating actively in grants. So that's been one of the challenges, is kind of bringing our board members up to speed on what does this mean. What does a partnership mean versus a grantee? And I think initially they all thought, "Well, it's a five-year commitment!" That's what a partnership means [in this case]: we will fund them for five years. And that's longer than we usually fund anything. So that was quite a step in the first place. But what it

really has meant is that more of my staff time has been involved [in helping make] decisions about the Coalition . . . in more of a leadership position.

The program officer added that it took time to convince her board that it was appropriate to assume a role as an active partner in the initiative. And she acknowledged that the "five-year commitment" is just the financial beginning. The goal is to keep the foundation active indefinitely as a funder and decision maker. While that may be a tough sell, the program officer concluded the following: "I think our board now has more ownership in it and more interest in the outcomes, and really can see the potential much more of it having a broader effect than just a one-time grant."

Sector-Deep Representation

As described in chapter 2, Representation might be viewed as the political demographics of the initiative. It is intended to capture descriptive aspects (who is involved) and the political essence of the initiatives, with an emphasis on those people and organizations that are commonly marginalized or uninvolved in youth development work. Representation involves questions such as: Who is represented in the new norm promoted by the initiatives? How widespread is the norm? Who has a voice in shaping it and in carrying it out? Is the norm applicable to all children and youth, and are the responsibilities for realizing this norm shared by all sectors of the community?

Within GivEm40, the Developmental Assets framework was integrated into the local school systems through its mapping onto the school improvement plan. Accordingly, representation in the initiative was rooted initially and deeply in the schools. A critical benefit to this approach is that all students who attend the schools are targeted for some level of involvement in the initiative. A potential risk is that the schools are situated to carry the early brunt of the work, setting up the potential for them to again be seen as overresponsible for the success and failure of the community's young people. A regional superintendent from one of the 19 school systems explained the integration of the asset framework into the school improvement planning process:

[E]very one of the schools has what they call a School Improvement Plan, which is written with specific goals and objectives in place that are supposed to improve student achievement. And it could be through improving your environment, it could be through changing classes or how you teach classes, or materials and supplies . . . We kind of jumped into the asset-building piece by having all our staff who are out in the schools working with professional development, teacher training, curriculum development . . . trained [in the asset framework]. [So] when they go and they talk about school improvement issues, we try to write in specific strategies that support the asset development into their goals . . . If I'm looking to improve your language arts score, for example, . . . there may be some things that are going to be developmentally appropriate from this asset list that we're going to fit into that strategy to help you, because we know that most of our kids don't feel real good about being in school, for whatever reason. We don't do a very good job of [addressing the larger] issues that impact [student] learning.

One Sector at a Time

Sector Connection is a central code within the Representation cluster; it captures the specific ways in which sectors connect to the initiative. The deep immersion into the school system allowed GivEm40 to carefully focus on a strategic approach for one particular sector—public education. The initiative leadership has used its school-specific approach to model a number of strategies for their subsequent work in other sectors. Key lessons learned within the public school sector include getting the key decision makers on board early, targeting specific assets around which to organize the work, and encouraging long-term involvement from one or more key funders. GivEm40 is using the careful planning of its public school sector to bring other sectors to the table, including the funding sector and the MSU research and evaluation team. By having a specific context with clear processes and outcomes on which to focus, the school sector can more clearly serve as a model for strategically involving other sectors. The GivEm40 team has been clear, however, that each sector will require its own processes of strategic care.

To exemplify the impact of the focused work in the school on the funding sector, a program officer of a small foundation stated, "People who apply for funds to support programs, we [tell] them, 'Show us what assets you're trying to reach.' We're following what Rotary Charities has done, what United Way has done." By organizing the Rotary Charities and United Way contributions around clearly defined asset building in the schools, the GivEm40 staff has been able to bring the asset language into the consciousness of the larger funding community, spawn monetary support for asset-building work, and pave the way for its efforts beyond the school sector.

Like the funding sector, the university involvement has taken an atypical path. MSU faculty members described the adaptations in engagement and mind-set required for their successful participation:

> [T]he shift we purposely made on this is that traditionally ... faculty come out and drive communities, and say, "You must do it this way." The approach we've taken here has been ... one of invitational interaction. We will respond when we're invited to respond ... And it's a real fine line, because what ... was not driving this is how many publications we can get out of this ... It's a very different format ... [than is typical for university involvement].

While publication opportunities for faculty might not be motivating the university's participation, the MSU researchers acknowledge other tangible benefits for their higher education community:

> For MSU it's been an incredible opportunity for us to do something we've really never done in quite this way from our land grant, which is trying to knit together our campus-based expertise and our field-based staff working with children, youth, and families.

In other words, the MSU research and evaluation team sees the GivEm40 initiative as a unique opportunity for connecting real-world youth development practice, and related research and evaluation needs, with the theory and research taught and practiced at the university. Accordingly, MSU's involvement in GivEm40 serves as a model for other colleges and universities in which faculty and students are seeking realistic ways of contributing to and learning from their communities, and doing so in a manner that is consistent with higher education's primary missions of teaching, learning, and conducting research.

Law enforcement has been a key sector for many of the HC • HY initiatives across the country. It is a sector that interacts with youth regularly, and often under the most deficit-based circumstances. As such, it is understandably one of the most resistant sectors or systems to adopting the Developmental Assets framework. In many initiatives, including GivEm40, police get involved in asset-based work largely through school-hosted programs. Because of its involvement in and subsequently beyond the schools in GivEm40, the police department systematically began working toward the adoption of an asset-based approach, with all the challenges inherent to that undertaking. The Traverse City chief of police was a self-described skeptic who ultimately became an asset champion:

> The hard part ... is getting [the asset approach] integrated within [the] services that we provide ... How do you deal with somebody [in a strength-based way] when you're putting the snatch on him? ... I mean when you take him into custody. [T]hat's a different approach than ... what we're used to ... Before it was very up front, very, you know, this is the way it happens. We have officers in our schools who have gone through the asset building and are trying to utilize those particular, different assets in their everyday day-to-day operation. Not only with youth ...

It's Simple, But ...

In contrast to the challenges police officers face in applying an asset-based approach in particular situations, simplicity was cited frequently as an attractive feature of asset-based work. But an interesting paradox emerged as we analyzed the data across the eight communities. The Developmental Assets framework was presented as extremely user-friendly, which was essential to its widespread adoption. Many of the actions required to build assets are simple and fit comfortably within everyone's capability to carry them out. But this simplicity also was described as deceptive. Although the actions might be simple, the attitudes required to activate them are more complex. According to respondents across all eight initiatives, many people are resistant to a positive orientation toward youth. Such resistance might exist at different layers of consciousness. For example, one can be superficially in favor of a strength-based approach while more deeply holding on to traditional beliefs that youth must be controlled and managed because of their immaturity, impulsivity, or developmental need to rebel against adult authority. According to GivEm40 staff, such engrained attitudes make it difficult to reach "critical mass" in their work. As a result, they put a great deal of energy into helping the community reach what they call a "blinding revelation of the obvious"—that youth will join adults when they are respected and valued rather than feared and controlled.

Because one can mistakenly assume that adults generally view youth from a strength-based perspective and are interested in nurturing those strengths, it is all too easy to underestimate how much time and energy are required to build the necessary support and infrastructure for carrying out the initiative's mission, according to GivEm40 staff:

> [T]hey [school leadership] thought [that] when we went to them with the Youth Friends [mentoring] program ... within three weeks a thousand people were going to be in that school ... And they were going to see huge results without giving some thought to the fact that we had to build the infrastructure in the school to make it effective ... We thought it was going to take a couple of months and it actually took about six ... When they replicated Youth Friends in Kansas, it took them 12 months before they had everything set up with their school districts ... Well ... I had 250 volunteers in three months ... but ... some of them had a problem with the schools because the schools weren't ready for them.

An interesting explanation for why it is hard to activate support for the work was offered by an initiative participant who cited the human tendency to forget our own developmental trajectories or pathways:

> And the beauty of our human nervous systems is that they're so efficient that we prune out all of that neurotransmission that got us to the obvious, and we forget how hard it was to move there. And so I think the difficulties just are perception, because we forget how difficult it was for us. To even be having this conversation denotes some level of sophistication ... [For] naive people coming into this arena, it would be really hard. But we forget, I think, how hard it was.

Beyond the Schools

Within its first two years of existence as an HC • HY initiative, GivEm40 seemed to be on its way to exemplary status as a model of community change that had incorporated the Developmental Assets framework. Keys to this apparent success included the grasp by initiative leaders of the connection between serious evaluation and funding sustainability. From the onset, GivEm40 secured its important evaluation partnership with MSU and its funding partnerships with the major foundations in the area, as well as the local branch of the United Way. It used a "top down" approach (e.g., recruiting superintendents prior to principals and teachers) in establishing receptivity and raising awareness among community leadership, prior to the launch of the initiative throughout the community. GivEm40 avoided the pitfall of spreading its efforts too wide. The focus was sector specific and deep, with its initial emphasis on the schools. Youth had been successfully engaged in leadership roles through the "Rally Around Youth" launch of the coalition, and 6 months later through the "Youth Summit on Violence Prevention." In short, GivEm40 was extremely strategic in its efforts to get out of the gates cleanly and strongly.

During our first site visit to Traverse City, the leadership was bracing for the public presentation of the second set of survey results. Although they knew that little change would be apparent in just two years, initiative leaders did not want to risk credibility with the community by changing the original plan of re-administering the survey after two years. By the time of our second visit several months later, the anxiety around the presentation had passed as the public seemed to understand that change would take time. During our second visit, a new concern had become primary. GivEm40 was now squarely focused on the need for a systemic change model to guide it beyond its sector-deep work in the schools. Leadership attributed the initiative's success in the schools partly to Clay Roberts, a Search Institute–affiliated consultant on implementation of the Developmental Assets framework within schools. They were now seeking "intermediate level" guidance for development of the business sector.

Search Institute was described as providing "very good lists of things individuals can do," but the research focused on the individual as the unit of change rather than larger systems. GivEm40 was seeking a model in which the community is the unit of change. From the perspective of the initiative's leadership team, Search Institute's unwillingness to take a clear lead in developing a deep and systematic community change model was a vital error, and a threat to the very survival of a Developmental Assets movement. The director stated during our second visit:

> My biggest frustration with Search is that it seems to be saying, "We're spawning a social movement," but does precious little to bridge that huge chasm of what researchers know and what practitioners do ... If we're model builders here [at GivEm40], then the world's really in trouble ... We're just groping our way through here.

Despite its vociferous call for systemic change leadership from Search Institute, GivEm40 had a clear plan for deeper engagement of the business community. At its core was the belief that "positive youth development" needed to be framed as "economic development." Businesses needed to view young people as important stakeholders in the community through their membership in the workforce and as consumers. Had youth development not already been identified on these grounds as a primary issue for local businesses, GivEm40 would not have turned its attention to developing this sector more deeply. Initiative participants argued that they did not have the staffing, time, or financial resources for the cultivation of early stages of mobilization to build assets; those resources needed to come from within the sector. The coalition's role was to promote the connection of "pivotal people" who enable the sector to progress with the Developmental Assets framework. Local businesses were motivated to play a role in "unleashing the potential of youth" as community residents and consumers in the present. A focus on thriving factors and the positive contributions youth seek to make in their community could spawn "youth friendly" businesses that would, in turn, encourage economic investment in the community and counter the export of youth from the region.

GivEm40 had also moved toward systemic change in the region through the leadership of individuals in the school district and family services. Despite its own references to the transformed "climates" of the schools through the leadership of Michael Kenney in the Intermediate School District, and Mary Marais in Leelanau Family Services, GivEm40 continued to seek an external road map for change at the systems level. The frustration of not feeling heard by Search Institute threatened to compromise communication even further. Search Institute was regarded as irresponsible for initiating the work but not supporting communities in sustaining their efforts with further community change research and planning. In fact, GivEm 40 staff challenged Search Institute to convene all of the "failed initiatives"—those that could not be sustained —at its next yearly conference.

The leadership of GivEm40 held a deep understanding of the complexities inherent in negotiating change within each sector of the community. Local idiosyncrasies, however, did not change the coalition's contention that there should be standard change models developed and promoted at the national level. Such standard models would then be adapted to local realities, according to this line of thought. As our study has shown, there are many blueprints for the successful adoption of the asset framework, with each community initiative bringing a unique context, structure, and orientation to its interpretation of the framework. GivEm40 clearly showed the makings of an emerging blueprint of its own. A key aspect of its approach was the willingness to take calculated risks at key junctures. Given the anxiety that inevitably accompanies risk taking, however, it is understandable that GivEm40 felt compelled to call out for help—to call for Search Institute to provide additional direction to help alleviate the need for risk taking at every turn.

Tenuousness and Survival

Feelings of tenuousness characterized all of the initiatives to varying degrees. Among our Wave I sites, Moorhead Healthy Community Initiative suffered threats of state funding cuts to its after-school programming. Orlando's Healthy Community Initiative was hit hard by the diminished travel and tourism following the terrorist attacks of September 2001. Portland's Take the Time staff were in the midst of reorganization following sudden changes in the political climate of the county commission that resulted in severe budget cuts, compromising the coalition's work and threatening its very existence. The feelings of tenuousness for GivEm40 were more conceptual in nature. Without access to existing models for promoting systemic growth, coalition members described the initiative as "groping along a pathway to change."

In the face of questionable financial sustainability, all of the initiatives have defined continuity of the work in broader ways. If funding "dried up," as some of the HC • HY leaders put it, the philosophy of Developmental Assets will have permeated their communities, and there was no returning to a pathologized perspective on youth. Some anticipated periods of initiative "dormancy" in which there was less visibility but ongoing behind-the-scenes influence. In the face of "death by reorganization," as staff of Portland's Take the Time put it, continuity was being defined in part through their participation in this case study, which served as one more means of making visible the pervasive, sometimes hidden, work of the initiative's partners.

A hallmark of GivEm40 was staff members' capacity to communicate and frame individual successes and failures as part of a larger process. Beginning with their own enthusiasm surrounding the Developmental Assets work to their more grounded experience in implementing it, they seem to have incorporated each step into their understanding and presentation of their long-term efforts. A participant in an initiative staff meeting expressed this fervor:

> I think we all . . . saw the Developmental Assets and saw this model and said, "Wow! This is so different. This is so new. This is so exciting and refreshing, that people are just going to grab it and run with it." . . . And so we really thought it was going to happen quickly.

Of course, that change didn't happen quickly. Mistakes were made along the way and looked upon as "lessons learned." An exemplary lesson came from what might be considered premature efforts to advertise the work before it was ready to go out:

> And we'd known not that much about public relations firms, and it just turned into this horrible relationship for both them and us . . . Part of it was really not understanding the full range of . . . their skills versus what we were asking them to do. Part of it was we really didn't know what we wanted to do, and that was changing. After working with them for about 10 months . . . we said, "This is really not working. Mutually let's just stop." And we'd found so much energy had been going into that relationship, trying to just work it through. (Initiative staff meeting participant)

As GivEm40 staff came to see it, they were focusing on survival and growth before they knew who and what they were. Although their preference was for more planful change, key turning points or critical events were thrust upon them from unanticipated directions. Their director's maternity leave marked one such event. GivEm40 staff described growing dependent on her in a host of ways, including her skills at facilitating complex communications. Suddenly, they were confronted with the need to develop better communication skills in her absence. The director, for her part, described feeling that she had become somewhat resistant to certain issues, and that her leave and the subsequent empowerment of the staff allowed her to work through her own areas for growth. The impact of key changes on staff resulted in an enhanced recognition that GivEm40 needed a realistic and robust strategic plan—one less dependent on the special skills of key individuals and one designed to take effect over the long haul:

> This idea of a committee of 15 and a staff of . . . at that point, three, magically transforming schools in six months, really, was silly. And plus we were learning a lot of things like even for people who really like this and grab on to it, it very much is a process and a continuum they move along. Even though you believe it . . . you still act . . . inconsistently with it. And so this notion that in six months these complex institutions with multiple goals are going to do this was silly. (Initiative staff meeting participant)

Spread Control

As initiatives move along their particular pathways of change, glitches of all types seem to prompt a stage of reassessment and taking stock, along the lines described earlier. All the initiatives appeared to pause at key junctures to assess their use of resources, including the efforts of committed citizens. In almost all cases, the initiatives assessed themselves as "spread too thin," and they began a process of limiting their expansion. Typically this process involved developing a shorter list of goals and a longer-range strategic plan.

Interestingly, there may be an inverse relationship between the degree of risk taking necessary for early stages of development and growth and the need for subsequent caution that prevents initiatives from spinning out of control. GivEm40 was characterized by both approaches: early risk taking and subsequent caution, control, and mindfulness with respect to strategic development.

Of the many lessons GivEm40 contributed to this case study, perhaps the most poignant is its recognition that going community-wide may be best accomplished, at least in some instances, by slowing down, by approaching the community one sector at a time, understanding the needs and leverage points within that sector, and then building from there. Our code *Spread Control* has its origins in GivEm40. It implies that when resources are too rapidly spread across the community, the impact is likely to be superficial and eventually fizzle out. When energies are harnessed, on the other hand, and used to drill deeply within a particular sector, those efforts are more likely to obtain sector loyalty. GivEm40 began where the kids are much of the time: in the schools. By going deep there, it set a tone of expectation for other sectors. To be part of GivEm40, one has to "walk the talk." It's not enough to give kids a little of each asset once in a while. To "GivEm40" in the Traverse Bay Area requires doing so "24.7."

Postscript

As a result of losing its financial support a few years ago, the GivEm40 24.7 coalition no longer exists, but its influence is evident in schools and youth-serving agencies in the Traverse Bay Area that are still using the Developmental Assets in their work. Some schools continue to use profile results to set goals. There are programs in the region's Catholic and public schools that use the Developmental Assets, including mentoring programs and a program targeting homeless youth. Michigan Works! Youth Works!, a nonprofit program that helps teens develop leadership skills and work experience, also integrates the assets in its work.

Chapter 4
The Forgotten Neighborhoods: Moorhead, Minnesota's Healthy Community Initiative

We follow the Traverse Bay initiative's emphasis on one particular sector with the focus in Moorhead, Minnesota, on the needs and invisibility of specific neighborhoods. In particular, concerns about underage drinking and lack of out-of-school activities among youth in some of the poorest White and Latino neighborhoods served as a focal point for much of this initiative's work. By collaborating closely with parks and recreation and community law enforcement, the initiative promoted such activities as a "Weed and Seed" program in an economically depressed neighborhood. The intent of this program was to stimulate neighborhood pride and a sense of civic responsibility among the youth. In addition to working with the parks and recreation and law enforcement sectors, Moorhead has pursued its neighborhood-based orientation by actively engaging elder leaders, supporting Latino youth as community leaders, and bringing the generations together to collectively build a stronger community.

Context of the Initiative

Moorhead began planning an asset-based positive youth development movement in 1994, following a visit from Search Institute's president, Peter Benson, in 1993. Moorhead Youth: MY neighborhood, MY community, and MY future, a precursor to the formal Healthy Communities • Healthy Youth (HC • HY) initiative, was launched in the community and schools with support and involvement of businesses, faith institutions, the education sector, youth-serving organizations, philanthropic foundations, city government, and law enforcement. To date, the most vital contribution resulting from the collective efforts of this group has been the organization and support of mentoring and after-school programs. One of the greatest challenges has been securing youth participation and leadership, including unsuccessful efforts to build and sustain a youth board for the initiative.

Moorhead Healthy Community Initiative (MHCI) began in 1994 in response to an increase in underage drinking and youth violence. It is the second-oldest Developmental Assets initiative in the country, following the establishment of the Children First initiative in St. Louis Park, Minnesota, in 1993 (see Chapter 7).

M.J. Nakkula et al., *Building Healthy Communities for Positive Youth Development*, The Search Institute Series on Developmentally Attentive Community and Society 7, DOI 10.1007/978-1-4419-5744-3_4, © Springer Science+Business Media, LLC 2010

Moorhead is a city of approximately 32,177 residents (2000 census). Although the vast majority of the population is White (approximately 94%), students of color represent 15% of the school-age population, with the largest minority group being Mexican American, followed by American Indian and African American. A primary employer is the sugar beet and sugar-refining industry, which has a seasonal migrant workforce. There are three institutions of higher education in the city: Concordia College, Minnesota State University Moorhead, and Northwest Technical College. Moorhead shares a border and a river with Fargo, North Dakota, a larger and somewhat more affluent community of approximately 70,000 residents. There is a history of barriers to partnership between the two cities, and Moorhead was described as the "poor stepsister," with a larger population of migrant workers and lower-income citizens.

Similar to Orlando's Healthy Community Initiative (Chapter 6), MHCI functions as a catalyst for youth development activity across the larger community. Whereas the Orlando initiative clearly adopted an identity as a guiding framework and stimulus for positive youth development activity in its area, MHCI was not as explicit about claiming its role as a catalyst, perhaps because MHCI provides a great deal of direct service in addition to functioning as an organizer or catalyst. The lack of explicit identification with and claiming of a publicly recognized catalyzing role contributed to a concern over continuity within MHCI. Its behind-the-scenes activity led to what some of the initiative leaders referred to as "perceived invisibility." There was a strong feeling that the initiative's role and contributions to the community had been lost sight of during a transition in leadership through which activities begun by MHCI were handed off to various community leaders. As described by MHCI participants, there was no communication system in place to promote this part of the initiative's identity—the role of a catalyst moving throughout the community supporting Developmental Assets start-up activities. Because MHCI is an active service provider, there had been little time for promoting its larger role of asset-building catalyst and getting the credit that was due. In seeking funding for the initiative, it was difficult to show concrete products or outcomes, since programs MHCI had begun were now functioning more independently and finding their own financial support. It seemed ironic that a community initiative that had been perhaps the strongest of those we studied in bringing visibility to the needs of low-income and minority youth, should now experience itself, 7 years later, as "invisible." In the words of the initiative director:

> Virtually none of those people [service recipients, funders, and other decision makers] know that we were involved. Because ... we didn't build in the structures that insist that ... our involvement ... be made evident to the people who are benefiting. So that's, you know, that's 5,000 people who may at some point be a natural constituency for us ... to raise funds, who ... ah, we'll not be able to tap.

One initiative staff member said the following:

> Parents just know that kids are able to stay after school and get homework help. They don't know ... sometimes they didn't know ... if it was Healthy Community [MHCI] that was funding it, or if it was the school district that was funding it.

Central to the context of MHCI is the pronounced need to enhance access to opportunities for isolated youth who had rarely been represented in community policy decisions. As the following statement from a focus group member indicates, access is a critical issue even as it applies to mainstream extracurricular activities:

> I think my favorite activity in this initiative is the money that we've been able to raise for kids. First it was youth of color, and now it's all youth who have, ah, their economic situation is such that they cannot afford to buy those pairs of skates to go out for hockey, or to pay for those piano lessons, to get involved in some kind of what I want to call an extracurricular activity. And you know, Moorhead is a big enough place that if you aren't practicing a golf shot at seven or eight, if you aren't playing basketball ... you're not going to be in any of those events in high school, in all likelihood. You can't show up as a ninth grader and say, "I think this is a golf club. Show me how to play golf," and get on the team, like you might in a real small town. And so to give those kids that same start, to me is something that I think is ... you gotta do that.

As noted earlier, there are three colleges or universities in Moorhead, and this higher education community represents an important resource for Moorhead, Fargo, and the surrounding area. According to MHCI staff, however, those youth who are isolated by poverty or racial segregation typically fall outside the influence of the higher education environment. As a focus group member put it,

> The new program that we've started, ah ... I called Diana once and I said, "You know, there's probably a lot of kids who just don't have access to a college campus. And for whatever reason, they may perceive that a college education or even just postsecondary isn't part of their story. You know, what can we do about that?" ... We created this program called "Linking Up" ... whereby every sixth grader got a field trip to one of the three higher ed institutions, over 250 sixth graders.

MHCI was driven by the lack of resources and opportunities for youth engagement in constructive use of time. There were few organized activities for youth, on the weekends and during the summers, and negative, racist stereotypes were also prevalent, according to focus group participants. Parents of youth from lower-income backgrounds were perceived as unaccountable. Lack of transportation limited access to existing youth programs, and peer pressure was viewed as having little competition from constructive activities and as having an increasingly negative influence. Among the HC • HY initiatives featured in this study, MHCI is characterized by its strong commitment to developing youth resources where none existed, and for assuring access to these resources for all youth, but particularly youth of color. One focus group member recalled the time

> ... when my son was trying with some of his friends ... to get a skateboard park going down there. Ah, one of the council people ... said, "Well, you know we don't do special interest groups in the park system." Somebody else said, "Well, what about the golfers, the tennis courts, all these other people?" I mean, what's the tradition of a park? A park is supposed to have swings, a swimming pool, tennis course, golf courses. Skateboard, what's that, you know? And the idea being that, well, traditionally kids are supposed to like tennis and golf. What's this little board with wheels on it? And ... to have an organization that says, "Well, maybe this generation has other ideas of what constitutes the way they spend their time, and that it might actually be healthy to have that to go to."

By addressing issues of access, isolation, and youth representation, MHCI has generated an impressive list of achievements, which includes the awarding of Youth of Color Scholarships, the establishment of Raices de Mexico After School Enrichment program, which travels and performs throughout the state, an American Indian Youth Leadership Camp, and the Quinceañera Club for preteen and teenage girls. All of these activities reflect a strong commitment to diversity, which has been a guiding theme in MHCI's work. From a socioeconomic perspective, after-school activities were funded by state grants to allow youth to attend free of charge. Barriers of access to the programming were addressed by increasing school bus transportation. As a result of their efforts, however, with the clear focus on representation and access for all youth, MHCI has at times been tagged with the criticism of only "want[ing] to help the bad kids."

Structural Features and Orientation

MHCI has had a large board of directors with 30 members who can serve up to two three-year terms. The size of the board is intended to ensure representation and a broad range of expertise. The option to serve two terms provides adequate time for building comfort with and understanding of the role and working context of the board, as well as to provide consistency and continuity to the work of MCHI. There are three paid staff positions, in addition to the director's position. In response to the concern over invisibility described earlier, a communications coordinator position was established to address the need to promote the initiative's work.

Focus on Out-of-School Time

The initiative's focus has been on developing programs and activities using the Developmental Assets framework to address the lack of positive youth development opportunities during out-of-school time. This effort has also engaged the school system, which often houses the before-school and after-school programming. The inclusion of community law enforcement is a distinguishing structural feature of MHCI, and the initiative initially shared a building with crime prevention officers of the community police force. In 2000, MHCI relocated from a downtown bank building to a former church rectory in a residential neighborhood. It is adjacent to two elementary schools, which has increased youth traffic between the schools, the after-school programming, and the MHCI offices, where several staff are involved in direct services. Most important, the new location has increased opportunities for positive contact between youth and law enforcement. Given a history of strained relations between minority communities and police in Moorhead, MHCI has been intentional in fostering more positive contact between law enforcement officers and youth of color.

MHCI is also linked to the university system through board members who are faculty in child development, education, counseling, social justice, and community partnership. Through this connection, Link Up, the mentoring program mentioned earlier, has used college students as mentors for young people in the community as part of a course requirement. Link Up is part of the national mentoring movement, and one of MHCI's roles is to serve as a mentoring clearinghouse in the area for organizations interested in starting mentoring programs or becoming mentors themselves.

New Director

At the time of our visits, MHCI was at a particularly interesting point in its history. Between our first and second site visits to Moorhead, Diana Hatfield, initiative director for 6 years, resigned, and Barry Nelson was hired. Nelson is a trained social worker and administrator from a prominent refugee agency, and he has a strong program and financial development background. He took our visit as an opportunity to better orient himself to the Developmental Assets model, which, he claimed, had not been emphasized during his hiring process. In addition to a new director, the board of directors was about to undergo major turnover as original board members reached the end of their second terms. In contrast to the feelings of discouragement in the initiative, reported by the field team in March 2003, MHCI seemed reenergized in October of that year.

Tension with Fargo and Partnership Potential

> North Dakota has been exporting its poor since it became a state. (Foundation officer)

A description of the structural features of MHCI must include Moorhead's conflicted relationship with Fargo, North Dakota, specifically in the early stages of the initiative's development. Based on what was described as "a long history of slights," initiative staff expressed concern that Fargo had no interest in genuine partnership, but would make superficial or token connections to Moorhead in order to secure recognition and funding. Sharing the "All American City" Award in 2000 is cited as an example of Fargo benefiting from the partnership but having little interest in Moorhead otherwise. One focus group participant stated the following:

> Fargo–Moorhead went together in 2000, to compete for All American City Award and they won. But the untold story is that in 1998 and 1999, Moorhead was already up there in the top 30 finalists, and we [MHCI] were there. And Moorhead Healthy Community Initiative was the [lead project] they were bragging about ... we worked our tails off, and we put in tons of hours to try to tell our story. The Civic League said we're all about regionalism now, so if you, Moorhead, really think you want to win, then you've got to do a joint application with Fargo.

MHCI members claim that initially a good deal of outreach had been made to Fargo and other neighboring communities in North Dakota to generate authentic partnership efforts on behalf of youth development, but with little success:

> Our superintendent and mayor have had two meetings with Dilworth, Fargo, and West Fargo to say ... "Kids don't stop at the river. Let's make this a Health Community region. Blah-blah-blah." We did our pitch. We da-da-da. They had absolutely no interest whatsoever in this. (Focus group member)

According to MHCI, Moorhead historically has been viewed as a "welfare magnet" by Fargo because of its higher population of migrants, low-income residents, and visible homelessness due to the lack of affordable housing. A focus group member described the *Fargo Forum*, the major newspaper serving both cities, as featuring negative news about Moorhead:

> We don't have our own newspaper, We don't get morning news from the *Fargo Forum* ... we're so doggone far from Minneapolis–St. Paul, even when they have legislative news [about Moorhead] in the paper, you look what it is: [it's] in this tiny graphic down there ... So Moorhead always ... the attitude is that we don't get our fair share of state revenues, [because we don't get the proper] attention.

Initiative members described concerns about losing representation if they were to partner with Fargo, owing to its larger size and disdainful attitude. However, they also recognized the movement toward regionalization, and, at the time of our second visit, appeared to be readying themselves for renewed efforts at building a trustworthy relationship between the two cities. Several past board members have been residents of Fargo, and the new director resided there, suggesting potential for cooperation. Indeed, in the development of the initiative since our two visits, a great deal of asset-building collaboration has occurred between the cities, as discussed in the concluding chapter.

As one of the oldest and most developed of the eight initiatives, MHCI is a unique contributor to this multisite case study. It was the only initiative to experience a change in directorship during the time of our study. Our visit was enhanced by the opportunity to view achievements, processes, and concerns through the eyes of the new director, who debriefed with us several times during the October two-day visit. Further, because Nelson was a new director, barely a month into the position, participants in our focus groups and related data-collection meetings had an added incentive to share information; they seemed invested in informing, and perhaps influencing, his leadership direction as well as providing our research team with information. The community presentation, during our return trip to Moorhead, had the largest audience of the initiatives to whom we presented our preliminary findings during our second trip to each site. That audience was particularly intent on sharing its views of MHCI and its wishes for the initiative's ongoing development.

Characterizing Themes

As touched on earlier in this discussion, dominant concerns of the Moorhead initiative had to do with funding sustainability and lack of visibility in relation to its efforts to create access to opportunities for youth from the most disenfranchised neighborhoods. The funding issues were intertwined with the lack of attribution to MHCI for changes made in the community on behalf of youth development. At the same time, however, the larger issue of financial sustainability suggested the need to encourage community ownership of the initiative and to locate funding from within the community, rather than feeling beleaguered by and estranged from funding sources that did not understand the long-term nature of the work. The initiative was striking, compared to the others, for committing its focus and finances so heavily to out-of-school time and youth access to resources. The focus on out-of-school time was successful in providing a range of opportunities for youth and in infusing the Developmental Assets approach and philosophy within the community. We wondered, however, whether a more balanced engagement with other sectors of the community—sectors beyond law enforcement and school- and community-based after-school programming—might contribute to broader and stronger community ownership of MHCI.

The various funding and sustainability themes outlined in the preceding section contributed to many of the codes for our study. In the remaining presentation of MHCI that follows, we organize these issues around two thematic clusters from our analysis: *Representation* and *Cultural Identity Development*. Whereas Representation largely focused on sector involvement in the GivEm40 analysis in Chapter 3, here it addresses the engagement of the community's demographic makeup, including ethnic groups and particular neighborhoods. Representation intersects with Cultural Identity Development in a number of respects. As the initiative attempts to involve particular groups through various activities, it strives to define itself—the very nature of its work, the visibility of its efforts, and the capacity to survive—as a vitally recognized contributor to the quality of community life. Visibility emerges as an organizing theme across the two thematic clusters: How to raise the visibility of the populations and neighborhoods that are frequently overlooked, and how to make the initiative visible as it seeps into the everyday workings of the larger community.

Cultural Identity Development

As with many of the other initiatives, MHCI experienced a poor fit between existing funding frameworks and its overall mission, resulting in a chronic tension between the core identity of the initiative and the need to keep it alive. As participants from various HC • HY initiatives have described, an obsession with financial survival can compete with and override the nature of the work itself, sometimes resulting in a sense of mission ambivalence and disintegration:

> So the challenge is that ... most of the funding sources are targeted to certain little issues, and we are broader than that.... And at times when we have nine different funding sources, it becomes a big challenge for Deb and [me] and our accountant and our treasurer to [put these resources together in a coherent package]. (Focus group member)

Ironically, MHCI's relative maturity as an HC • HY initiative fostered a heightened experience of disconnection between standard funding approaches and the Developmental Assets framework. After the initiative had been in existence for a number of years, early funders were leaving as they typically do following initial investments. Perspectives on timing constitute perhaps the most obvious disparity between common funding approaches and the needs of the HC • HY initiatives. Programs are typically funded on a yearly basis, or at best on a two- to three-year basis, and are expected to produce measurable results in an unrealistically short time frame. Initiative leaders know that they are creating a longer-term vision and related approaches that must be maintained and measured over time in order to show meaningful change. An MHCI director's report referred to the 20–30-year time frame applied to other social movements or initiatives, such as seatbelt use, MADD (Mothers Against Drunk Driving), and the antismoking campaign. Adopting a similar time frame, and procuring adequate levels of funding, would give MHCI and other HC • HY initiatives a chance to produce impressive results as well, according to the initiative leaders. But that vision is hard to realize when funders are only willing to make short-term investments.

Adapting Funding Frameworks

Our basic code *Unique Adapt* captures the need for each local HC • HY initiative to adjust the asset framework to its unique community characteristics. But just as the initiative must adapt to local conditions, there is also encouragement for particular community sectors to adapt their usual ways of doing business to become more congruent with the initiative's mission. MHCI, like some of the other initiatives, presents a clear example of encouraging philanthropic organizations to adapt their traditional funding schemes to the needs of the initiative. Specifically, such organizations are asked to become part of the initiative rather than simply investors in it.

Funding challenges within each HC • HY initiative are inextricably linked to the nature of partnerships, which create the core of each initiative's cultural identity. These partnerships highlight a critical tension between traditional funding approaches and the HC • HY mission, which is characterized by sharing resources and bringing disparate constituencies together to forge an environment fostering healthy youth development. Funding decisions based on requests for proposals often foster competition for scarce resources and, as a result, a splintering of the youth service community. This mismatch between funding approaches and the HC • HY mission raises the question of whether initiatives should concentrate on engaging funding organizations to become part of the initiative, rather than operating as an external provider of revenue, thereby fundamentally altering the culture of fundraising and the culture of the initiative itself. In several of the initiatives, including

MHCI, we heard foundation program officers describe the appeal of working with the Developmental Assets model, and the potential openness to altering common funding strategies. One officer reported the following:

> In our early days, we were very much a general purpose grant maker. We had all kinds of grant applications coming in for this year's Campfire program or Boy Scout program, but very little that had any kind of long-term impact. So we started looking for models ... and ... we modeled a grant program, really, on the kind of things that were being done in Moorhead ... with the Healthy Community Initiative. ... And for the last five years, we've been building that program in partnership. We're a funny funder in that we don't just hand people money and go away. ... We tend to ... kind of get in partnership with the people ... to whom we provide resources. And so we've learned a tremendous amount from Moorhead. Moorhead's benefited by [our] money, but I think, also, by the assistance of our staff at various times with various issues ... their staff might be facing. So it's been ... a real two-way street for us.

As these comments suggest, some progress has been made in bringing funders on board as initiative partners. But that progress too often has been the exception. There are substantial obstacles to funding organizations becoming comfortable in partnering with particular entities, even if they are community-wide collaborative endeavors such as the HC • HY initiatives. One concern is that an initiative like MHCI might become a funding clearinghouse, distributing resources across affiliated programs and in effect serving as another funding resource for competing programs or service sectors. Another concern is that philanthropic organizations by their nature need to remain viable resources for a host of competing organizations and approaches. By linking too strongly with one approach or another, the foundation runs the risk of losing its autonomy and independence. These concerns create real limitations in the degree to which philanthropic organizations can uniquely adapt to the long-term needs and interests of the HC • HY initiatives.

A Guiding Story of Financing Innovation

Another strategy for addressing the incompatibility between traditional funding approaches and the goals of the initiative was to encourage community ownership of the initiative's financial health, much as a congregation or college approaches financial development through donor programs and planned giving:

> Well, the problem with that [traditional foundation funding] then is that everybody's focused on that area where the pool of resources is rather finite, versus the capacity of individuals ... of individual giving, which is much less restrictive and finite. People's generosity is, frankly, the biggest, you know, resource we've got out there. (Focus group participant)

The decision of businessman Steve Scheels to initiate a matching grant as an incentive to encourage local businesses to contribute to the Moorhead initiative serves as a model for stimulating community ownership. Scheels had given large grants over several years without the matching challenge. His revised funding strategy created a vehicle for his personal giving to leverage the contributions of friends and colleagues, and in that sense broaden the funding infrastructure of the initiative. Funding approaches like this one shape the very nature or culture of the initiative.

Sheels's initiation of a matching-funds approach became an example of our basic code *Guiding Stories*, in that it served as a model for thinking about MHCI's sustainability and community ownership. If funding came from individuals within the community as well as long-term foundation partners, there would be a sense of local ownership, not only in the work of the initiative, but in its fiscal health as well. The more the community could own the initiative, the more it could strategically shape its identity, rather than having that identity determined by the next pot of money and whatever strings might be attached to it.

Branding and the Complexities of Recognition

The development of each initiative's identity is inherently linked with the specific contributions it makes and with the credit it receives for making those contributions. This simple formula holds a two-part challenge: identifying the contributions and then associating the initiative with those contributions:

> "Do these things actually happen?" You've gotten the money, "Great, look at all the numbers, look at all these [things you've done] . . . but what assets?" I'm not saying that they haven't happened, but it has been a concern in the past, saying, "Okay, you said you did this. What kind of outcomes?" How do you measure, just because somebody says, "Yes, I interacted with an adult." Is that an asset? (Focus group participant)

Capturing the tangible results of the asset framework convincingly enough to sustain funding is a central challenge in all of the initiatives. Because MHCI serves as a catalyst, clearinghouse, and community brainstorming center, in addition to providing direct services, assessing its impact is particularly challenging. In some cases, the initiative had been the catalyst for starting or seeding a program that subsequently went on to secure funding independently of MHCI. One such example of this was the Quinceañera Club for young women In this case, the initiative seeded the program financially, and it provided valuable networking information for staffing, implementation, and the securing of continued financial support. Two years later, the club became independent of MHCI's funding, and there was no longer a concrete connection between the initiative and the program in terms of contribution and recognition. Is the Quinceañera Club part of the culture of MHCI? It is for those who recognize the historical connection. But for those who don't have the benefit of that perspective, the club might be seen as a more progressive or culturally relevant opportunity than MHCI for certain youth. Such losses of recognition contribute to fractures in initiative identity, which over time can lead to exacerbated funding crises.

Another example of illusive attribution was the initiative's fostering of mentoring programs. MentorLink is an MHCI program linking mentors from Concordia College with elementary youth at a local public school. In addition to implementing their own mentoring program, MHCI provides training and support for other mentor program providers as an active participant in the national mentoring campaign. MHCI staff described receiving many calls from mentoring programs and organizations seeking advice on funding resources. They also made important networking connections between organizations that would go on to implement their own mentoring programs. Translating this type of influence into impact that could be assessed according to funding criteria was felt to be beyond the initiative's expertise.

A related challenge to capturing impact was the difficulty of teasing out the contribution made by MHCI when it had partnered with other organizations or institutions. For example, as a result of interacting with MHCI in some capacity, several faculty members in area universities had incorporated the Developmental Assets framework into their curricula in teacher, counselor, and social justice education. How would one go about capturing the impact of indirectly contributing to the professional development of future teachers and counselors? If MHCI was no longer connected to a successful program, the contribution seemed abstract, and not easily apparent, even if it had served an instrumental and necessary function and made a significant contribution. To help address this dilemma, MHCI hired a communications coordinator. One of her functions has been to promote recognition of the role of the initiative in the community, but even recognition falls short of assessing impact, which is the real challenge.

Branding is one of the essential concepts central to the Cultural Identity Development cluster. This concept refers to the importance of creating an identifiable set of markers that helps make and keep the initiative visible. Ideally, the brand would be placed on or associated with the key contributions made by the initiative. Although this notion can appear superficial or self-aggrandizing, it is critical to keeping the initiative alive in the public eye. Leaders from most of the HC • HY initiatives touched on this point repeatedly in our interviews. While the ideal might be to catalyze work across the community, to share ownership for healthy youth development as fully as possible, there is also a need to remain a definable, recognizable entity if that entity is to serve its catalyzing function. Branding, in one form or another, is a means of doing just that.

Lack of Sector Diversification and "Providing for . . . "

Each initiative gives a good deal of thought to how it expands to reach more youth and involve more sectors. Our code *Processes of Expansion* captures the various ways initiatives foster the growth of their work. For MHCI, the commitment to providing opportunities for youth was so pervasive that it risked truncating the initiative's capacity to diversify its efforts to build collaborations with a diverse range of sectors. MHCI led the development of an extensive after-school program, along with a morning gym component, and mentoring activities to address idle, unsupervised time on the weekends. The tremendous amount of time needed for building this direct service programming dramatically reduced the time remaining for coordinating development efforts with stakeholders in other sectors.

Paradoxically, the intense focus on youth service provision and program development might inadvertently have led to a reduction in youth leadership of the initiative.

There was some indication that youth involvement in the running of MHCI had declined. As we will see in Chapter 4, the decision by the Healthy Community Initiative (HCI) of Greater Orlando not to focus on the schools and not to be a direct service provider created more time for diversifying its efforts at sector recruitment and youth leadership cultivation. Orlando's successful engagement of youth through a philanthropy initiative indicated high expectations for young people and trust in their capabilities.

Our study of the eight initiatives suggests that struggles engaging youth may require a reexamination of the nature of youth involvement. A heavy dose of service provision for youth at the cost of engaging them in initiative leadership can have long-term consequences related to authentic affiliation. If youth cannot identify the entity with which they're involved, they will not be able to serve as ambassadors and recruiters within the larger youth community. This comes back to the issue of branding discussed earlier, but is also indicative of a basic code that emerged frequently in our analyses: *Doing With, Not Providing For*. This concept represents the tenet that working with youth as partners is preferable to providing services for them, in that the former stance is empowering whereas the latter one reinforces power differentials.

The cultural identity of MHCI seemed to be at a crossroads. Lots of youth were being served by the initiative's efforts, but relatively few were guiding those efforts. The initiative was highly successful in creating opportunities for youth in the education and recreation arenas, but relatively few community sectors were being cultivated as active partners. MHCI's leadership spoke clearly about these dilemmas and was in the midst of thoughtful planning at the time of our data collection. The outcome of that planning will have been critical to shaping the MHCI's culture going forward, including its effectiveness at including a wide range of youth, adults, and sectors of the Moorhead community.

Representation

> Listening means that what is heard gets taken seriously and considered when decisions are made. (Focus group member)

The challenge of establishing a youth center has been a continuous theme in the life of MHCI. Many youth seem to have grown frustrated by this history of disappointment. Young people in Moorhead have consistently voiced the desire for an age-segregated center, without younger children or seniors, which might be interpreted as inconsistent to some extent with the Developmental Assets focus on fostering intergenerational contexts. Adult staff members within the initiative have emphasized this potential inconsistency in their interactions with the youth. While the logic might make sense in theory, however, it doesn't hold up in practice, according to many of the teenage participants. They argue that without a place for teens to congregate as a distinct group, many young people simply will not show up at all.

To intensify this difference in opinion, MHCI youth had raised money through various activities and proposed the creation of a juice bar. Adults expressed concern that older youth, who might have been drinking alcohol elsewhere, would then mingle with younger peers at the proposed juice bar, reinforcing the problem of underage drinking that is so pervasive in Moorhead.

It is clear from these differing and even contradictory scenarios that questions regarding function, location, and staffing of the center are far more complex than failed attempts at funding and might help explain why a center has not been built. It was unclear how these concerns had been processed with the youth. In "Moorhead Speaks ... Community Listens," a community-wide assessment of MHCI, youth criticized decisions made without their input or without attempts to negotiate different points of view, suggesting the need for more time and training spent on authentic adult–youth dialogue.

Representation, the second thematic cluster featured in this chapter, represents the demographic and political involvement in the initiatives. For Moorhead, representation was very much an issue of youth presence and voice. It is ironic that MHCI, which so passionately worked to create access to education and recreation opportunities for youth, struggled just as hard to authentically involve a diverse range of youth in its decision making.

A Strong Start in Building Representation

In 1993, the superintendent of schools invited Search Institute's president, Peter Benson, to Moorhead. The community was introduced to the Developmental Assets as a framework based on research showing that an increase in the number of assets youth experience in their lives could counter negative and risky behaviors. Community leaders, such as the chief of police, the mayor, university faculty, leaders of congregations, and businesspeople were in attendance, indicating Moorhead's readiness, as a community, for an alternative perspective on addressing their concerns about youth. Asset Teams, involving over 300 volunteers, were mobilized and held town meetings throughout the community to raise awareness of the Developmental Assets framework and its application in specific areas. The "After School Hours" Asset Team conducted a survey of youth to learn their interests for after-school activities. The "Neighborhoods Asset" Team organized free trainings in Developmental Assets for neighborhood block clubs, in partnership with the Moorhead Police Department. The "Intergenerational Relationships" Team led to what is now the MentorLink program.

The move from volunteer to hired staff with a significant operational budget was viewed as a significant turning point in the initiative's history. Between 1994 and 1996, Diana Hatfield was hired as the initiative director, the board of directors was established, and the Search Institute *Profiles of Student Life: Attitudes and Behaviors* survey was administered to students in grades 6–12. Initiative members stated that broad representation of all community constituencies was highly intentional and critical to activating the initiative. In 1996, Moorhead was selected as the pilot site for the state after-school enrichment program, and MHCI was designated as the organization best equipped to implement the programming based on its operationalization of the Developmental Assets framework. Grants that were awarded locally during that period reflected a changed school climate and activities that had not previously existed. A high school teacher described the changed climate in the schools, as a place where the emphasis had been on "getting youth out of the building." Now the after-school programming was expanded to morning activities, such as allowing students to use the gym before school.

All of this early success signaled that MHCI would likely be a national model for the Developmental Assets movement, and in many respects that has proved to be the case. But such successes come with costs. As the initiative was deemed successful in designing and implementing services, it became somewhat locked into that role, contributing to the struggles in diversifying youth and sector involvement.

Forgotten Neighborhoods, Forgotten Youth

In 1996, Moorhead was selected as one of four sites for Minnesota's "Weed and Seed" program, and the city chose to focus on the Greenwood Mobile Home Park, where Joe and Cory Bennett, two long-term residents and Developmental Assets advocates, had brought attention to the needs of low-income youth. A partnership with the Parks and Recreation Department was undertaken to upgrade park equipment and supplies for youth in the "forgotten" neighborhoods, such as Greenwood. Promoting the work through his yearly Christmas party, Joe managed to secure upgraded park equipment, sneakers for his soccer team, and bike helmets for neighborhood youth. He has spoken before the state legislature on behalf of the neighborhood's needs numerous times, and it seemed that his testimony was finally directing attention to some of Moorhead's neediest communities.

The commitment of the community policing agency to such neighborhoods was demonstrated by the Police Department's continuing the funding of neighborhood-based positions when the "Weed and Seed" money ended. A local university also became an established partner in the forgotten neighborhoods through the advocacy of faculty who had been active on the Asset Teams and were now on the MHCI Board. Minnesota State University Moorhead professor Steve Grineski's students were matched with youth in four neighborhoods, including Greenwood, for community-based tutoring, which functioned as preservice professional teacher training. Mentoring activities were bringing college students to the neighborhoods and changing the "town vs. gown" nature of the university, where students remained on campus. Specialized after-school programming for Latino youth was developed and championed through the efforts of Outreach coordinator Sonia Hohnadel, and $13,000 was awarded through Youth of Color and Thompson Scholarships in 1 year. Raices de Mexico, a performing arts program, promoted a positive image of diverse youth through statewide performances.

People like the Bennetts and Steve Grineski are referred to as *Key Bridges* in our study—influential people who make connections across sectors or bring isolated parts of communities into relationship with other people, institutions, and activities. The Bennetts helped put Greenwood Mobile Home Park on the Moorhead map in a different way, and in doing so brought a range of resources to the young people living there. Similarly, by directing his students to interact with children and youth in a range of communities, Steve Grineski helped make the university more tangible to young people, just as it made those communities real places within the university. Such bridging across sectors is at the heart of the representation cluster and underscores the critical importance of key people in leveraging the resources and good will of the larger community.

Are We Doing What We Set Out to Do?

In 1999, the *Profiles of Student Life: Attitudes and Behaviors* survey was conducted for a second time with youth in grades 6–12, with disappointing results. The average number of assets for youth remained unchanged. The lack of assessed change, following 5 years of hard work, prompted the return to a grassroots assessment modality in the form of the "Community Speaks ... MHCI Listens" project—a study conducted by researchers at Minnesota State University Moorhead and longtime initiative participants Brenda Shafer and Nancy Frosaker-Johnson. Between February and April 2000, Shafer and Frosaker-Johnson conducted 16 focus groups, with 126 participants, to review the results of the survey and to learn the community's perspective on future direction. Four of the focus groups conducted were with high school–age youth, who had been active in the Youth Board of MHCI. One board member noted that "the initiative should be this connecting force and focusing its efforts on linking youth services and organizations rather than on providing specific services."

The "Community Speaks" project ushered in a reassessment phase for the initiative. The project provided important opportunities for reflection and dialogue among the adults and youth in the community. During this period, the community feedback suggested that there needed to be more diversification in sector outreach. Respondents argued, for example, that businesses needed to support flexible hours and family time as a positive value countering the "frenzied lifestyle" of many families, and that religious congregations needed to explore further opportunities to move away from age segregation in church activities. The feedback was strong that efforts to build Developmental Assets needed to focus on adults as well as youth. Specific areas targeted by the study were adults' need of parenting education, particularly with older youth, since there was adequate support for early childhood parenting.

Another area addressed in the feedback was the need for better adult role modeling to counter the drinking culture of Moorhead. The need for more opportunities for youth and adults to interact meaningfully was cited as essential to building credibility among youth. The failure to successfully negotiate building the Teen Center, where youth could safely enjoy "down time," was again cited as a central concern. Overall, youth expressed the need to feel that their input was seriously considered in decision-making processes, particularly in areas where there was adult–youth disagreement. While it might seem like mixed messages resulted from the "Community Speaks" project (there needs to be more adult–youth interaction, and teens need more time for themselves as a group), the overarching message was quite clear: Teens want more of an authentic voice in decision making. That voice, it seems, would argue for meaningful intergenerational connections as well as safe spaces for teens to spend time among themselves with as little adult monitoring as possible.

But whatever activities were to be undertaken, another message stemming from the "Community Speaks" forum was the importance of thinking through the nature of asset building within the activities. In the words of one focus group member:

> Okay, we gotta have an after-school program. Would the community come back and say it's gotta have assets explicitly built in? The question is, so how does this advance the building of assets in Moorhead youth? ... [A]nd that's the organizer, that's the thing that gives it some kind of a real important dimension.

During our second trip to Moorhead, we learned that the after-school funding had been under siege for not explicitly addressing youth remediation issues, such as academic problems. There were serious questions about the fate of the programming if it needed to become a volunteer-run effort again because of lost funding. When we raised this concern with the staff, the response was that the Developmental Assets approach had permeated the community and that "there is no turning back" to deficit-oriented approaches to youth development. This response prompts further thinking about the definition of "continuity" within these initiatives. Are there periods of dormancy, or less visibility, during which the initiative still continues to exert a strong influence, philosophically and otherwise? It seems to be the case that for an HC • HY initiative to remain an effective and visible entity within a community, it needs powerful advocates of its philosophical center. When money dries up during certain periods, at least for particular activities, how does the initiative sustain its energy? How does it hold the course when the easier route is to chase the new money, and be pulled in the direction of the funding stipulations, whether or not they are consistent with the HC • HY mission? These are the issues embedded in MHCI's question from the "Community Speaks" forum: "Are we doing what we set out to do?" And if not, "how do we get back on course?"

The Element of Risk Taking

Each initiative contributed uniquely to our understanding of effective leadership—a theme we coded *Leadership Determination*. HCI of Orlando contributed to this theme through its relentless emphasis on the importance of local, grassroots-distributed leadership. By raising awareness of the degree to which leadership is representative of constituencies that are not traditionally represented—forgotten neighborhoods and isolated groups of youth, for example—MHCI took the courageous risk of investing its energies where the needs were greatest and the political clout least present. Unfortunately, by initiating services and activities in these areas, and then turning the leadership over to local constituencies, MHCI placed itself at risk of being forgotten or rendered invisible. Given these dual risk factors, it is critical to recognize the model being implemented here: It is a model that at once addresses the core of Developmental Assets work—building on the strengths of all youth, including those typically left behind—and one that emphasizes distributing responsibility for this work throughout the community. While this model is consistent with the heart of the HC • HY initiative, it pushes it to the edge of invisibility, precisely where it should be, arguably, and precisely where it is most difficult to be sustained owing to lack of recognition. This is an issue that needs to be taken up seriously if the HC • HY movement itself is to be sustained.

Postscript

In November 2006, MHCI was part of a panel presentation on community-based solutions at the third annual George Sinner Symposium on Public Policy held in Moorhead. Out of that presentation, a steering committee composed of leaders from Moorhead, Fargo, and West Fargo was formed to examine the possibility of a metrowide healthy community movement.

In May of 2007, Search Institute's student profile survey was administered to 17,000 students in grades 4 through 12, in all public and private schools in the metro area. The results of the survey were released to the public in a community-wide event, "Join the Wave for Metro Youth," held on September 25, 2007. A new organization, Metro Youth Partnership (MYP), has been formed with a new and expanded board representing the larger community being elected in January of 2008. The community moves forward now, committed to creating "the best place on the planet to be a kid."

Chapter 5
Pursuing "The Tipping Point": Portland, Oregon's Take the Time Initiative

Whereas the first two case studies we presented (Traverse Bay, Michigan's GivEm40 and Moorhead, Minnesota's Healthy Community Initiative) focus on particular sectors and neighborhoods, Portland's Take the Time is organized largely around two interrelated theories of community change: diffusion of innovation theory, as articulated by Everett Rogers, and social threshold theory as discussed in Malcolm Gladwell's popular 2000 book *The Tipping Point*. Diffusion of innovation theory, which argues, in part, that influential early adopters of new ideas are critical to persuading others to try them out, is used to foster a grassroots orientation to mobilizing the community. Specifically, Take the Time initially used the widespread awarding of mini-grants to get the Developmental Assets framework in the hands of a number of potentially key early adopters who might persuade others to follow their lead. This strategy was designed to stimulate local ownership of the initiative and to reflect the value placed on individual contribution, creativity, and egalitarianism.

By organizing local involvement in this manner, the initiative sought to attain enough early momentum for asset-based activity to effectively counter what its leaders described as the negative portrayals of youth promulgated by contemporary society and reflected through the media, both locally and nationally. Reaching the "tipping point," however, requires that exemplars of successful grassroots activity effectively link with key leaders and community decision makers who are essential to helping integrate new ideas into the very fabric of the community. Take the Time worked strategically to build these connections, but learned how difficult it is to work broadly, at a grassroots level, while simultaneously trying to influence powerful institutions such as the media to support a coherent change of direction in their treatment and portrayal of youth.

Context of the Initiative

The Commission on Children, Families, and Community of Multnomah County (CCFC) launched the Take the Time initiative on October 29, 1997. According to a 2000–2001 evaluation report on Take the Time, CCFC develops public policy, conducts planning, and undertakes initiatives to promote thriving families

M.J. Nakkula et al., *Building Healthy Communities for Positive Youth Development*, The Search Institute Series on Developmentally Attentive Community and Society 7, DOI 10.1007/978-1-4419-5744-3_5, and communities. Take the Time has been clear from the outset that its work does not constitute a program. Rather, this countywide initiative operates as a catalyst, building on existing efforts and infrastructures that already support positive relationships between adults and youth. Working from this starting point, the initiative relies explicitly on three strategies to promote growth that would ultimately reach a tipping point whereby the goals of the initiative would become the common pursuit of the larger community:

- Support individual asset builders with practical assistance;
- Provide technical assistance within targeted systems; and
- Conduct mass communication.

Take the Time uses these strategies to work toward the following three fundamental goals:

- Children and youth have a relationship with a caring adult;
- Children and youth have a meaningful role in the community; and
- Children and youth are valued and supported by their community.

"Death by Reorganization"

Feelings of tenuousness characterized all eight initiatives in the case study to varying degrees. When we interviewed Take the Time staff, they were fighting for survival in the face of changes in the county political environment, which in turn resulted in a replacement of the initiative director. Changes in the leadership of the county commission had resulted in different funding priorities and severe budget cuts that compromise the work of Take the Time. The dedication of initiative staff to the Developmental Assets movement was compelling and evident in their strong desire to share "lessons learned" with us in this study in the event their particular initiative did not survive. Amidst what the staff described as a harrowing previous six months of "paralysis" due to funding uncertainty, and attacks on the integrity of the initiative, the director and her staff reminded one another that "this was not about us" and committed themselves to a plan that would continue the work on behalf of children even if the initiative were disbanded.

In the face of tenuous financial circumstances, all of the initiatives have defined continuity of the work in broader ways. If funding "dried up," it was hoped that the philosophy of developmental assets would have permeated these communities to such a degree that the return to a deficit-based perspective on youth was not likely. Some leaders anticipated periods of initiative "dormancy" during which there was less activity and visibility, even though the initiative would continue to exert a strong influence through its established emphasis on an asset-based approach to supporting youth. In the face of "death by reorganization," as the Take the Time staff put it, the initiative leadership defined "continuity," in part, through the connection established with Search Institute and the Developmental Assets movement by means of this case study. Whatever the future of their particular initiative, it was hoped that their contribution to the knowledge base would help embolden and energize the work of others. Initiative staff members pursued that goal vigorously in their day-to-day work through attempts to influence the media and through their support of local service providers who applied a strength-based perspective.

As alluded to earlier, Take the Time was started as a campaign within the Commission on Children, Families, and Community of Multnomah County. As a county-funded initiative, it is subject to the instability of shifting political tides. The emergence of a new county commissioner resulted in the replacement of the executive director of Take the Time in August 2001. A succession of severe budget cuts to the social marketing campaign, a key structural feature of the initiative, limited strategic messaging to the community. Further, because of the civic mandates of the commission, the managing director's efforts to restore support for the initiative were criticized as exercises in self-interest designed to save her job.

Multnomah County has 615,000 residents and includes the city of Portland, surrounding suburbs, and small, outlying rural communities. During the early 1990s, a statewide network of citizen-led, local commissions supporting healthy community development was initiated through the governor's office. The intention was to mobilize and empower local communities to develop conditions supportive of positive youth development. This shift was a dramatic move away from an emphasis on professionalization and pathology, toward supporting community potential to promote the healthy development of youth. Since established indicators of youth health were largely deficit based, such as statistics on school dropout and pregnancy rates, and substance abuse arrests, Multnomah County sought a tool for defining and measuring wellness. Search Institute's *Profiles of Student Life: Attitudes and Behaviors* survey provided a measure of protective factors that was consistent with the emerging community philosophy. Take the Time was launched in response to findings from the survey, which had been administered to 10,000 students in grades 6, 8, and 10. Results indicated that Portland's youth were struggling to receive the assets needed for healthy development.

Structural Features and Orientation

Take the Time is the only initiative we studied that was situated within a local government structure. Staffing included a managing director, outreach coordinator, grants coordinator, communications director, and clerical assistant. A steering committee composed of 18 youth and adult volunteers was active in conceptualizing and implementing the work. At the time of our study, Take the Time had recently expanded to include four more counties—a change the leadership had been reluctant to make previously owing to the lack of a clear working plan. To share the initiative's name, Washington, Clackamas, Marian, and Clark counties agreed to adopt the three-pronged focus of Take the Time's strategic plan. Multnomah County provided technical assistance to the new county members. The move to expand was similar in some respects to the decision made by the Moorhead Healthy Community Initiative (see Chapter 4) in relationship to the city of Fargo, North Dakota. In both cases, expansion was viewed as perhaps the only way to continue the initiative's work.

For Portland and the surrounding counties, a major strategy for mobilizing the community around Developmental Assets was the awarding of "mini-grants"—small grants of several hundred dollars, given to more than two hundred organizations and individuals. This strategy served multiple purposes: It spread the application of the work and its message throughout the community; it established a wide network of asset builders and people committed to developmental well-being; and perhaps most important, given the tenuousness of the political climate, it planted seeds that would be especially important should the initiative not survive. Take the Time's work within the school system, for example, was designed to support school-community partnerships by helping seed them with mini-grants. While the monetary sums were small, they planted ideas among students and provided them with modest resources with which to plan. From a diffusion of innovation perspective, the mini-grants promoted and expanded innovation; they encouraged youth and adults to bring their ideas and energy into the mix.

The Developmental Assets framework was not viewed by the Portland initiative as *the model* of community change but rather as "the what and the why," with the "what" representing a strength-based focus and the "why" representing the present and future health of the larger community. As discussed earlier in the chapter, diffusion of innovation theory served as a guiding model for the initiative. The theory argues that change is disseminated person by person, and that successful innovations are characterized by simplicity, observability, apparent advantage, user friendliness, and consistency with existing values. These factors are organized into five stages of individual change, postulated by diffusion of innovation theory. The first stage, Knowledge, emphasizes accessing vital information necessary to making change. This is followed by Persuasion, which addresses utilization of the knowledge from stage 1 to encourage making change. Stage 3, Decision Making, emphasizes commitments made to address the issues requiring change. Stage 4, Action, focuses on the specific activities undertaken to bring about the targeted change. And finally, Sustainability addresses the conditions that must be put in place to maintain and build upon the accomplishments achieved. Among the initiatives we studied, Take the Time was unique for its clear focus on strategies and goals that were aligned with a guiding theoretical model. In the sections that follow we show how their structural and theoretical focus contributed to the actual work and helped contend with the challenges confronted.

Characterizing Themes

Take the Time provides a vivid portrait of the need to start a change process within each individual and to link that individual growth across the community to reach a workable consensus on what can and must be done on behalf of a common mission. The stories from Portland's youth and adults show inspiring examples of individual reflection and growth, juxtaposed with accounts of how challenging such actions can be for some people. Perhaps more than anything, Take the Time teaches us that while community sectors are vital to the efforts of Healthy Communities • Healthy Youth (HC • HY) initiatives, real people make up those sectors, and as a result achieving meaningful results requires reaching real people.

Personal Ownership

Take the Time is an interpersonally driven campaign, which might be expected from an initiative interpreting innovation as spreading through relationships among people. Relationships are seen as "the key to getting anything done." The initiative's work is imbued with explicit relational values—trust, valuing individual contribution, openness to learning, and equality. The work emphasizes listening, rather than "preaching," the Developmental Assets, and honoring the individual, program, or sector's existing system—a facet of the *Unique Adapt* code, which we found among all the initiatives. According to the director, "One very important principle of how we implement all of this, is to try to make it relevant to an individual or an organization's unique situation and their own agenda. So trying to fit it to their agenda, instead of saying, 'You have to take on our agenda.'"

Participants describe their involvement in Take the Time as personally gratifying, but also requiring a level of introspection and self-assessment of personal biases and beliefs. Again, in the director's words:

> I think the Youth Advisory Board taught me a really important lesson . . .which was my own ingrained ageism, and my recollections of how when I graduated from high school, I had all these ideas, and then graduated from college and had all these ideas . . .I felt like everybody told me, "You don't matter until you have some experience, and we don't want to hear what you have to say." So I spent a whole lot of my twenties trying to gain the experience to have a voice. Then came this Youth Advisory Board. I thought, Well, it's not their turn. You know, you have to wait like I did. And suddenly I thought, This is really stupid. It's wrong. And I had a very profound awakening, and I realized just how much we as a society miss by doing that to young people, by not wanting to hear what they have to say.

This passage links two basic codes within the *Personal Ownership* cluster: *Baggage Assessment* and *Awakening*. Time and again participants across the initiatives echoed the sentiments of Take the Time's director. Effectively carrying out strength-based youth development work often requires an assessment of the ways in which one has internalized deficit-based attitudes toward youth and how that influences one's interactions with them. That assessment process is commonly triggered by an awakening of sorts, by the lightbulb suddenly going on and shining an uncomplimentary glow over the hidden baggage of our own development. The next step of constructively challenging these newly recognized beliefs and prejudices is much more difficult, but it is required to experience a genuine awakening of the type described by the participants in our study.

Once awakened, it is critical that we *live out* the commitments to strength-based development in all aspects of daily life, according to HC • HY participants from Portland and elsewhere:

> I think we all have thought in every arena that we operate, whether it's our own families or in our work [that we have a responsibility to model the assets], and I . . .on my way over here my children called me to say, "This is really funny, that you're going to talk about taking the time . . .two nights in a row!" So, whether it's within our own families or within our own . . .professional [lives, we have that responsibility] . . .and I think that that's one thing we have talked about a lot on the steering committee, how we live our own lives, how we will model what we are doing in every area . . .whether it's professionally or personally. (Staff meeting focus group)

The basic code *Living It Out*, captured in this passage, closely relates to another code within the Personal Ownership cluster. *Owning It* depicts the importance of applying the asset framework to oneself, of following the process of assessing one's "baggage" described earlier with a deeper knowledge of how to build strengths in oneself as well as with and for others. The transformation of our communities, from this perspective, is dependent on a prior self-transformation, on taking ownership of one's own role in the current state of affairs and in a process of change that begins with personal responsibility. According to a focus group participant:

> Well, when you talk about transformation that, of course, is a change in attitude. But I think you have to get people to the point where they're recognizing their own attitudes first, and really saying, "What is my attitude toward youth? Am I afraid of them? Am I willing to meet them on an equal basis?" We need to begin to tap into what their attitude is. Then you have the ability to change it, and go into transformation.

As we listened to this focus group member more closely, it was clear that the "they" she referred to is each of us. Community change begins with self-change. Like voting, each individual action—each act of self-change—is essential to the workings of the larger process. But putting the individual pieces together into an integrated working model is a tall order. Our thematic cluster *Reaching a Common Ground* captures part of the challenge.

Reaching a Common Ground

Bringing together adults and youth from different backgrounds, who hold different points of view, and who can contribute different resources is central to reaching common ground regarding how to carry out the work of asset building and community transformation. Reaching common ground is somewhat different from reaching consensus, at least in certain respects. Differences in approach and opinion were viewed as healthy by Take the Time as well as the other initiatives, as long as there was a common *guiding purpose* with an explicit focus on strength-based work. The common ground appeared to serve as a solid foundation for the work, a stable place philosophically from which to debate, brainstorm, plan, and revise the implementation of the Developmental Assets framework. Without such grounding, differences risk splitting coalitions apart. With it, according to the participants in our study, there is a legitimate chance of building and sustaining diverse constituencies whose members can bring their passions together for a common good:

> ...Like me and Carson, under a normal situation, I probably wouldn't go up to Carson and talk to him because I just, like, we're not exactly alike. We don't ...we have some of the same interests, but we're just, we're not the same ...And there's some other [relationships on the] Youth Advisory Board [that are] even more crazier than our connection, but it's like, it allows me to talk to who I wouldn't usually. And I always find the person that I least think that I'll get along with, or mesh well with, is like the person that I can work so well with.

This sentiment expressed by a 16-year-old participant in Take the Time vividly captures the role of what we have termed *dissimilarity* in our analysis. There is a unique form of energy that comes from the dynamics of difference. New ideas are often integrations of existing differences of opinion; new implementation strategies are often syntheses of differing ways of doing things. Because the Developmental Assets movement is conceived of as a community-wide undertaking, there is an assumption of reconciling difference at its core, an assumption that differing people and approaches can and must be brought together to strengthen communities for all young people. While this assumption seemed to hold across all eight communities in our study, the negotiation and integration of difference, at all levels, were simultaneously driving forces for making productive change and particularly demanding challenges to bringing about that change.

Egalitarian Context

In Chapter 2 we discussed the *Context Create* theme that was common to all eight initiatives as the need for a physical and philosophical "space" for locating the efforts and exchanging ideas on behalf of positive youth development. Take the Time participants described "creating an atmosphere of high regard" for individual contributions and new ideas. The intent was to make this space as egalitarian as possible so that youth, and adults who might not otherwise participate, would find a generative home in the work. Among the eight initiatives studied, Take the Time was perhaps the most understated in terms of formal leadership, though strong leadership was readily apparent. The following statements from different adult staff members among the leadership team reflect components of the initiative's avowed values:

> It's about trust and valuing people's individual contributions. That we don't have all the answers. That we're here together equally. So those are kind of underlying values ...that I attach to Take the Time.

> It's a matter of creating an atmosphere ...an atmosphere that builds trust, [where people] value each other.

> [It involves] contacting somebody and really focusing the first meeting [with them] around building a relationship, learning their stories, sharing some of your own. And really leaving it there, not trying to push an agenda ...But using relationship, which is, I think, a key theme.

Spreading the Word: Successes and Setbacks

Consistent with the egalitarian approach of valuing the individual creativity of youth and adults, Take the Time used the mini-grant approach introduced earlier to mobilize the community as broadly as possible. At the time of our study, over $100,000 had been awarded in small grants, ranging from $200 to $500, for approximately 200 asset-building projects and activities. The wide range of grants awarded emphasizes breadth over depth—as contrasted, for example, with GivEm40's focus on the regional school systems. Examples of mini-grant activities include a white-water rafting trip for grade 8 girls, designed to address assets 16: High Expectations, 30: Responsibility, and 15: Positive Peer Influence. A regional program for the deaf and hard of hearing was funded to provide sign language books for parents of deaf toddlers, addressing asset 2: Positive Family Communication. Portland public schools were awarded funds to plan activities for an African American leadership conference, which would address assets 32: Planning and Decision Making; 8: Youth as Resources; and 3: Other Adult Relationships. Community gardens, community cleanups, reading buddies, foster grandparents encouraging pre-reading skills, intergenerational quilting, and Saturday tutoring for students in grade 3 are other examples of the wide range of asset-based activities that were supported through the mini-grant mechanism.

The mini-grants were successful in helping raise awareness about Developmental Assets to the extent that 883 proposals were reviewed over the course of three years. Accordingly, even the projects that did not receive mini-grants were brought into the orbit of the initiative, thereby allowing them to learn more about initiative's work and to network with potential partners. Another core dissemination strategy for Take the Time was the use of media campaigns. The initiative used a wide range of advertising media and promotional strategies to spread the word, including billboard and bus signage, radio public service announcements, television news coverage, print news, movie screen advertisements, and development of a Web site. Despite these efforts to use the media strategically, the promotion of Take the Time's launch was experienced as a disappointing failure, as expressed by the following focus group participants:

> It was our big launch and we thought it was going to be that everyone was going to come. And you know, it didn't happen . . .the rest of the community didn't kind of go, "Wow! We get it."

> The hired promotional consultant failed to deliver the visibility through planned media coverage, and the initiative suffered a serious blow to its credibility from that mistake . . .It was one of the biggest setbacks . . .of the whole campaign. Not just because of the event itself, but actually because of what it did to the relationships between staff and the commission.

> Because we had built expectations so high . . .the commissioners were like, "This is not what you sold us . . .Where is this stuff [sophisticated media campaign] that was supposed to happen?" This was a huge failure. I think we all went through a very hard phase for about four months, where we were just stuck with that and had to really process through it . . .If there was ever a time we thought about giving up, that was the time.

There is some question as to whether Take the Time ever recovered fully from that initial setback. Internally, the staff described working through the disappointment and stigma of "failure." Despite the failed launch, the initiative's work continued, and Take the Time went on to achieve important successes through the mini-grant process, a dramatically improved media campaign, and related community work. Nonetheless, the coalition was left with the lingering feeling that a major opportunity—perhaps *the* major opportunity—for political support was irrevocably lost.

Youth Advocacy for Balanced Media Coverage

> My survey found that 1 in 14 stories in the daily sections was about people ages 12–22. More than 4 in 10 of those stories were about college or high school related sports, almost 2 in 10 were about crime, and nearly 2 in 10 about youth-related political and social problems. Fewer than 1 in 10 stories was about accomplishments by young people. ("Youth Isn't Being Served as Well as It Should Be," *Oregonian*, September 12, 1999)

We have described the paradoxical nature of asset work as seemingly simple yet too infrequently done, and we have underscored the tremendous efforts required to shift perspectives on youth from deficit to strength based. Both of these issues apply fully to Take the Time's efforts to influence media coverage of youth. It would seem simple enough for the media to write positive, uplifting stories of young people, yet this is infrequently done. Indeed, shifting the media's focus from youth problems to youth accomplishments, particularly outside the athletic arena, proves more challenging than it would seem. Bad news sells, particularly dramatic bad news such as brutal crimes, which produce an emotionally jarring impact. And this impact, of course, brings with it a lasting residue of negative feelings toward youth, who are often featured in these dramatized depictions.

Convincing the *Oregonian*, the major regional newspaper, to research the claim that news coverage of youth in Portland was primarily negative is a tremendous achievement. While Take the Time is understandably proud of having pulled off this accomplishment, its next steps were even more newsworthy. As a result of the initiative's lobbying, the *Oregonian* hired a writer to expand its depiction of youth. The writer's job was to seek out stories of young people's accomplishments, stories that cast Portland's youth in a different light—one more representative of the full range of who they are.

To take its efforts to influence the media even further, Take the Time facilitated youth involvement in the Journalism Credibility Project of the American Society of Newspaper Editors. Through this project, Portland's youth advocated for more accurate representation of young people as a form of journalistic credibility. In a sense, the goal of the media efforts was to reintroduce Portland and the larger country to its youth, to a truer depiction of them. Perhaps the most powerful way to make this reintroduction, however, was to let youth speak for themselves. It's one thing to influence adult-generated reporting on youth; it's quite another to have youth speak up for themselves.

Young people's involvement in and ongoing contribution to creating "The Zone," a weekly front-page feature on youth in the Friday Living section of the *Oregonian*, is a strong example of authentic and proactive engagement. More specifically, it is a stunning example of the contribution youth themselves can make to countering pervasive fear on the part of adults, and educating them about the strong capacities of youth to contribute to their communities.

In reflecting on the media efforts undertaken by youth and adults, Take the Time staff looked inward for validation of the need to do this work:

> And I think that the initiative has pushed us in some ways . . .Some of it is just reinforcing simple, everyday things that are very, you know, greeting a child . . .But it [also] pushed us . . .to really think about youth involvement—it's one thing to just sort of say, "Well, for the sake of the grant . . ." It really pushed us to be very thoughtful about what that [youth involvement] meant, and involving youth in a way that wasn't just bringing them to the adult table and asking for their opinion . . .but really integrating them into this project. (Focus group participant)

"The Zone" epitomizes the work done within Take the Time to ensure that youth involvement is genuine. The *Reaching a Common Ground* cluster includes a number of essential concepts that capture these efforts. The code *Evidence of Credibility* suggests that the larger community and its various constituencies need solid evidence that the work is having an impact. The articles in "The Zone" provide such credibility. They show the larger public that Portland's youth have a voice and are committed to one another and to improving their community. *Levels of Engagement* asks about the depth of involvement in the initiative's vision. Both adult and youth immersion in the media project reflects the level of engagement needed to ultimately reach common ground regarding adult perceptions of youth. Finally, *Peer to Peer* captures the importance of youth reaching out to each other on behalf of the initiative's mission. There were few examples of this code in our study as poignant as youth reaching youth through Take the Time's media focus.

Parent Outreach

All of the initiatives express the need for outreach targeted toward parents, and in general, their efforts have proved challenging and frustrating. Some have appeared to "shelve" these efforts in favor of focusing on the school environment, where they have access to youth, teachers, and school administrators. Take the Time's Middle School Outreach Project suggests that although progress is slow, small but significant inroads can be made in this area. The initiative hired a Middle School Outreach Project coordinator to develop and maintain a network of parent outreach organizers in 13 middle schools. Each parent organizer attended monthly three-hour trainings with other parent organizers from across the network, and scheduled two or three events annually at their school to increase parents' and guardians' understanding of how to help their children be successful students and community members. Trainings focused on research on the Developmental Assets, parenting skills, and adolescent development. A core aspect of the trainings centered on encouraging parent communication with school personnel and raising their awareness of school policy advisory opportunities. The activities of and receptivity to the parent outreach efforts varied widely at each school. Parent organizers used newsletters, "family nights," "recognition rallies" for youth achievement, and school visits for students in grade 5 transitioning to middle school and their parents.

Through an evaluation of the parent organizers' efforts, Take the Time learned that parents struggle with understanding the Developmental Assets. This struggle points to a critical barrier to reaching a common ground: If we don't speak a common language, it is difficult to formulate a common mission and pursue common goals. A second finding from the evaluation of the parent organizers points to a potential solution: Parent organizers were found to be most successful in reaching parents when there was relatively little turnover in the role. In other words, when there's an opportunity to build understanding over time, it's easier to get on the same page. A somewhat unanticipated finding of the evaluation was the impact parent organizers were found to have as role models, not only for their children but also for other parents. They were often working full time but still managing to actively participate in their children's schools. These parents were not only organizing other parents but also serving as models of community involvement for them.

Two of the basic concepts from the Reaching a Common Ground cluster capture the parent work in Portland. Although *Peer to Peer* refers to youth involvement in our analytic scheme, it is applicable to parents as well. Parents are critical to one another in this work. It is one thing to be encouraged by program directors and even school administrators; it is quite another thing to be encouraged and supported by fellow parents taking on similar challenges. In this regard, our code *Shared Learning* applies to the benefits parents glean from each other. Shared child-rearing stories, for example, are reassuring whether they capture celebrated successes or nagging challenges.

Community Change: Person by Person, Mistake by Mistake

The value of the individual as change agent is a common thread woven throughout Take the Time. All community members are seen as potential contributors, with wide efforts made to reach and involve them, as exemplified by the minigrant process. In their exuberance to build an inclusive coalition, the initiative staff acknowledges making mistakes of inexperience and naïveté:

> I think we all . . .saw the Developmental Assets and saw this model and said, "Wow! This is so different. This is so new. This is so exciting and refreshing, that people are just going to grab it and run with it." . . .And so we really thought it was going to happen quickly. (Focus group staff member)

In their efforts to reach a common ground, Take the Time staff do not blame their mistakes on others or chalk them up to ill fortune. They accept their "failures" and recast them as "lessons learned." Similar to Traverse Bay's GivEm40 initiative (Chapter 3), Take the Time hired a consulting firm to help guide its media approach and advertising strategies, and as with GivEm40, the consultation proved ineffective. Rather than dwell on this failed attempt, Take the Time interpreted the experience as a statement on its own expertise.

Ending the consulting relationship freed up energy for other matters, and, in fact, that step seems related to Take the Time's becoming its own public relations expert. None of the other seven initiatives in our study so doggedly pursued the media and related communications outlets in their efforts to spread the word about positive youth development. Individual and collective mistakes led to new and deeper learning and solidified the initiative's direction.

Turning points or critical events in the life of the initiative are defined interpersonally, such as the director's maternity leave. Staff described having been dependent on her to facilitate communication and suddenly needing to develop better communication skills with one another in her absence. The director, for her part, described feeling that she had become somewhat resistant regarding certain issues, such as the initiative logo, and that her leave and the subsequent empowerment of the staff allowed her to work through that resistance. This reflexive stance allowed the staff to recognize the need for a more realistic strategic plan. A focus group member shared the following realization:

> [T]his idea of a committee of 15–20 and a staff of, at that point, three, magically transforming schools in six months, really, was silly . . .Plus we were learning a lot of things, [such as] even for people who really like this and grab on to it, it very much is a process and a continuum they move along . . .Even though you believe it [the asset approach], you still act inconsistently with it. And so this notion that in six months these complex institutions with multiple goals are going to do this was silly.

A staff retreat provided the opportunity for reassessment, gathering new information, and deciding to shift direction. A similar process was described by most of the eight HC • HY initiatives in our study. As the initiatives became fully engaged in the activities they had planned, there was commonly a period of reassessment and recognition of the need to target a narrower range of activity. We labeled this phenomenon *Spread Control* in our analysis, based on the common description of the range of activities spreading out of control and compromising the effectiveness and reputation of the larger mission. For Take the Time, this reassessment process resulted in the three goals and three strategies listed at the beginning of this chapter. Fostering authentic youth involvement is at the heart of these goals and strategies.

Our study of the eight HC • HY initiatives is very much a study of youth representation. For such representation to move beyond token presence on committees with adults, a significant and ongoing orientation shift for adults seems necessary, even for those adults who might be working most closely with youth. Take the Time helped us identify this need most clearly owing to the staff's transparent acknowledgment of their deep-seated biases. "Walking the talk," according to Take the Time, requires relinquishing control, building and earning trust, and taking calculated risks. In many respects, engaging youth in partnership requires more than providing services and activities for them. It requires authentic relationships and all that they entail. And much of what such relationships involve is the confrontation of our own deficit-based biases. Perhaps more than anything, Take the Time has taught us that community change not only starts with self-change but also requires a constant commitment to it. When adult growth is restricted, so too, by definition, are the possibilities for healthy adult–youth relationships.

Postscript

Although budget cuts ended Take the Time as an initiative, the messages of the initiative—that children need to be valued by the community and sustained by adult attention—endure in homes, schools, child-serving organizations, and the actions of individuals throughout the community.

Chapter 6
Community Sustainability: Orlando's Healthy Community Initiative

In an area of exceptionally high mobility, Orlando's Healthy Communities • Healthy Youth (HC • HY) initiative focuses on building a sense of community coherence and ownership by emphasizing community revitalization activities. Examples of this approach include "listening projects," which consist of interviews conducted by youth and young adults with neighborhood residents to learn what they envision for their neighborhoods, and what is needed to make that happen. Youth then bring knowledge from these interviews to youth philanthropy planning meetings. By learning to assess the needs of the community and linking them with planning grants, youth are mentored into civic leadership roles. Through extensive efforts to connect residents of all ages in their community revitalization efforts, the Healthy Community • Healthy Youth Initiative of Central Florida (hereafter referred to as HCI or the Orlando initiative, consistent with the initiative's self-referencing) hopes to encourage more of its citizens to "put down roots" rather than just passing through.

Founded in 1998, the Orlando initiative focuses primarily on four broad areas:

- Children and Families: The "Crusade for Children" has the goal of making Central Florida the best place in the world to raise a family and to be a child;
- Community Connectedness: Building community trusteeship through growing social capital;
- Diversity: Promoting the acceptance and appreciation of all members of the community; and
- Center for Trusteeship: Establishing a neutral community place for the free exchange of ideas. Schools, businesses, congregations, youth and family service organizations, and city government are all viewed as part of the center.

Context of the Initiative

When HCI was founded, approximately 500 people moved into central Florida on a daily basis, and roughly 300 of them will have moved out over the next 5 years (*Orlando Sentinel*, November 20, 2001). High population mobility in the greater

M.J. Nakkula et al., *Building Healthy Communities for Positive Youth Development*, The Search Institute Series on Developmentally Attentive Community and Society 7, DOI 10.1007/978-1-4419-5744-3_6, © Springer Science+Business Media, LLC 2010

Orlando area has resulted in the depletion of natural resources, such as tree canopy for housing and roads, air quality, lake systems, and the unique species of flora and fauna that are the very reasons people are so attracted to the area. Further, the heavy mobility into and out of Orlando means that people do not "put down roots" and become invested in the preservation and sustainability of the community, both ecologically and economically. Not only does the vast tourism industry burden the ecology with its worker turnover, but the service economy that supports tourism has a similar turnover rate. In central Florida, 45% of all jobs are in the service industry, 13% in retail, and 7% in manufacturing (*Orlando Sentinel*, November 20, 2001). The terrorist attacks of 9/11 crippled the economy of central Florida in the early part of the decade, owing to its heavy dependence on the travel and tourism industry, and, at the time of our site visits, the recovery process was proving difficult at best. Heavy dependence on tourism is seen by HCI as a fragile and undiversified economic foundation for central Florida, which prior to Disney's arrival in 1971, depended on agriculture and manufacturing.

It is within this context of a fragile economy and high mobility that positive youth development is viewed as seriously compromised. Forty-four percent of elementary school children do not finish a school year in one school (Legacy 2000, p. 30), and child poverty is higher now than it was 10 years ago (Legacy 2000). HCI materials highlight studies showing high mobility related to low performance on school outcome measures. Related to the high mobility that affects school readiness is the need for affordable housing that might interest renters in becoming home buyers. The move from tenant to home owner is seen as a political priority, and as such our visits to Orlando included tours of several community revitalization efforts featuring new affordable housing options. An understanding of the economic, social, and ecological context within which HCI operates is critical to understanding the initiative's unique application of the Developmental Assets framework, and the ways in which community sustainability is linked to developing youth as active community members in the present and for the future.

Structural Features and Orientation

HCI is located in the city of Orlando, which in the 2000 census had a population of approximately 500,000 people. Greater Orlando is one of nine distinct areas or broad communities within Orange County, which in 2000 had a population of 977,000. Five of the nine areas—University, Winter Park, Maitland, West Orange County, and Apopka—are part of the HCI network. Ray Larsen, the executive director of HCI, is a former director of the Beta Center, which provides transitional housing for pregnant minors. Larsen has published in the areas of school readiness and child wellness, which helps explain some of HCI's priorities under his leadership. HCI's office is located in a former neighborhood health clinic building, further illustrating the wellness orientation of this initiative.

Early funding of HCI came from two hospital systems: Orlando Regional and Florida Hospital. Florida Hospital no longer provided funding at the time of our study, but still donated use of the building rent-free. HCI's beginnings also included substantial financial support from the Winter Park Health Foundation. In addition to youth health needs, the focus areas of the foundation include access to health care for all residents, and issues specific to seniors, such as "end of life care," lifelong learning, and driver safety. The foundation makes explicit links between the needs of teens and seniors, such as improved nutrition and increased physical activity, and it uses the Developmental Assets framework in grant making. Although a health and wellness orientation is the organizing theme for HCI, an overreliance on the financial support of the health care industry would seem to reinforce HCI's economic vulnerability.

The *Legacy Report*, an HCI publication, is the product of an assessment tool, the Compass Index, which is used to measure community well-being and help guide HCI's work. A brief description of the index illustrates the nested nature of Developmental Assets within the larger mission of community sustainability. The Compass Index, developed by Orlando consultant Art Kissen, evolved from a model of community wellness known as Daly's Pyramid. In this model, the local economy rests on a foundation of the natural resources and ecosystem that are unique to the area. Economic production, in turn, is the foundation for the development of social systems and culture, such as education, government, social services, health care, and the arts. At the top of the pyramid is the well-being of the community, which is determined by the perception of residents' health, happiness, and personal fulfillment. Kissen's Compass Index changes the hierarchical nature of Daly's Pyramid to the four directional quadrants of a compass, with "social, economic, well-being, and nature" corresponding to north, south, east, and west, respectively. This modification implies that each of these quadrants affects the others, and that no one quadrant is either foundational or the ultimate goal. Community members' well-being, for example, can lead to caring for the natural resources of an area, just as those natural resources can support well-being.

Characterizing Themes

As with all eight HC • HY initiatives in our study, the vast majority of the codes from our overall analysis can be readily found in the Orlando data. In this presentation, however, we feature the thematic clusters *Synergistic Commitment* and *Leadership Wisdom* because of HCI's unique contribution to these aspects of our analysis. In fact, it is the connection between these two clusters that makes the Orlando initiative particularly interesting. As one reads the remainder of the HCI story, a guiding question might be this: What types of leadership are optimal for maximizing synergistic commitments to healthy youth development across the community?

Synergistic Commitment: Initiatives within an Initiative

HCI is perhaps the purest illustration of our basic code *Context Create*, which we found to be common to all eight HC • HY initiatives and central to the essence of their work. *Context Create* refers to participants' descriptions of the need for a philosophical and physical space in which to locate the work of the initiative. HCI stresses neighborhood and community ownership of programming, and sees itself as a catalyst and *holding environment* that is composed of many initiatives. As a catalyst, HCI instigates action across the community; as a holding environment, it creates a space for that action to be integrated with other efforts and mutually understood among a broad array of constituencies. Larsen has described the process as "rebuilding community at the meta-level and also at the very local level." By "meta-level," he is referring to the global community of caring and responsible citizenship, as well as the broader regional community of central Florida, whereas by "local level" Larsen means the individual institutions and neighborhoods that make up a particular community.

HCI is distinct from the Orlando HC • HY initiative in Larsen's view. He sees HCI as the vehicle for community revitalization and sustainability, and HC • HY as a specific tool, or "lever," used within the larger mission of developing and sustaining a robust, engaged community to counter the impact of high mobility in and out of central Florida. In other words, there are two interrelated missions at work in Orlando: the specific community revitalization and sustainability mission of HCI, and the youth-specific community development mission of Orlando's HC • HY initiative. Although the distinction might seem like splitting hairs, it captures a reality that exists across a number of the initiatives in our study and beyond. The HC • HY national movement, which spawned the local initiatives, is commonly incorporated within existing community structures. At times those structures become absorbed within the local HC • HY initiative; at other times, the reverse is true. Orlando views its approach as an example of the latter scenario.

To make matters yet more complicated, HCI defines discrete activities or groupings of related activities as their own initiatives, and links them to the HCI mission of community revitalization and sustainability as well as to the HC • HY mission of community-wide youth development. The unit size or scope of these smaller initiatives is anything but uniform. Such initiatives might include a school cluster composed of a high school and feeder middle and elementary schools. A neighborhood working to organize support for a new charter school or youth club would constitute another example. Within these subcommunity initiatives, the Developmental Assets provide an overarching language and strength-based mission that fosters a sense of cohesion, or as Larsen has described it in an interview:

> What I mean by [metalevel community revitalization] ... is that the local community owns, in a sense, that subcommunity's initiative and by owning it I mean they are not only the ones to plan, but ... they [also] got to find the local resources ... They've got to find a partner ... and the strategy is, if a meteor comes and hits my organization, Winter Park and West Orange will continue with their initiative. So it's in effect self-sustaining and very locally owned. At the same time we work hard to be kind of the center that holds and so we have a community asset network. We have a Web site. We facilitate. We connect. We do all the things so that overall there is a strategy ... a cohesive strategy for the community as a whole, but the way that we're going about it is by nurturing these kind of sub-communities in building assets.

Larsen goes on to describe the importance of local geography in conceptualizing and developing the notion of subcommunity initiatives:

> Another strategy that we've used ... is that we've divided our community, at least on one level, into geographic groups. They're called cells and by that we mean ... naturally forming [generating] subcommunities—[the] community that surrounds a high school and the feeder schools, middle schools and elementary schools that surround it, [for example] ... That creates a cohesive geographic area. It creates a place where most people in the area share services, share churches, grocery stores ... So it's kind of [its own] community. And we build capacity within the subcommunities and then facilitate them perceiving how they want to [use the] assets in their community.

In a community like Orlando, or central Florida more broadly, it is particularly understandable that a great deal of emphasis would be placed on geographically defined subcommunity initiatives. If mobility in the larger region is high, that global concern might feel too large to manage effectively. If, on the other hand, smaller subcommunities can create conditions that foster "putting down roots," as Larsen aptly articulated the goal, more global revitalization and sustainability might be accomplished one initiative at a time. It is precisely from this line of reasoning that the code *Context Create* was developed, and the larger thematic cluster *Synergistic Commitment* created. If a context can be created to nurture the discrete initiatives from an overarching perspective, facilitated by a common language, the synergistic impact of these partnerships is likely to be greater than the sum of its parts.

Not Slamming the Schools

Although school engagement is considered a prerequisite for building community sustainability at the sub- and larger-community levels, schools themselves are not the focus of the Orlando initiative. In fact, HCI intentionally sought to mobilize in sectors that were not the schools, or in subcommunities surrounding schools, given the typical indictment of the school system as the source of all that is wrong with "youth today." As two focus group participants put it:

> You know, these kids feel their communities have let them down, not their schools ... We need to put our children with the community, and not rely on the schools to raise our kids. It's just not fair.

> We want to shove schools ... down there [away from the center of everything] ... We ... like them [school personnel] to [see it as] a safety net—mentors, faith institutions, or other adult relationships. And then the schools ... can be allowed to teach.

HCI staff expressed empathy for school administrators and teachers who often feel that new programs or approaches are suddenly thrust upon them to address the litany of existing "student problems." That expression of empathy was appreciated by those school personnel who are actively involved in HCI. One teacher shared:

> As a school system we're used to people saying, you know, "Here's all 40, now go do them." And it's been ... real helpful ... to say that we are just a piece of that, and the community has to come together and fill in the rest of the pieces.

But convincing teachers that "this is different," that they are not going to be left carrying the ball alone, has proved to be no easy task, as an HCI staff member who presents the HC • HY initiative to potential school partners indicated:

> You have to understand how beleaguered our teachers are. And I cannot emphasize that enough . . . [E]very time we [say], "This is [a] community [initiative], this is not a program for the school," but I [know] I [am] going to walk into a room full of teachers basically saying, "Yeah, right, something new from central office."

The task, according to this staff member, is to help teachers understand that "this is not about what you have to do," but, rather, "about what you are already doing" as it relates to the asset framework. *Not a School Slam* is a code we developed to capture this perspective, which is epitomized in Orlando but evident elsewhere as well. This concept, with its emphasis on the role of nonschool sectors in the community, highlights an important tension within the HC • HY work. Some initiatives, like GivEm40 in northern Michigan (see Chapter 3), focus deeply on the school sector as a means of organizing support structures for educators and students, not as a means of pathologizing them. Others, such as HCI, view even a heavy focus on supporting schools as implicit blame, or at least an overemphasis of their role in holistic youth development. A key concern associated with a heavy school emphasis is that it can enable the lack of sector diversification across the community. It is tempting to go where most youth spend most of their waking hours. HCI argues that that temptation can lead to community stagnation. The creative work, from their perspective, resides in the intersections of the subcommunity initiatives, whether those be neighborhood organizations that include schools, or sector collaborations that center on economic and physical wellness, for example.

Virtual Communities

HCI cautions against the assumption that community sectors such as education, religion/faith, business, or health care function cohesively, or share communication even within the confines of their specific arenas. The initiative leadership argues that references to sectors as "communities" implies a common language and shared goals, an assumed reality that often does not exist. It may be misleading, for example, to think of the business sector as a cohesive community. Accordingly, it may be just as important to foster community development within sectors as it is to promote it across them. From this perspective, Ray Larsen argues, the notion of a "faith community" or "education community" should be considered a "virtual" concept. Sector-defined notions of community are virtual communities in that they exist in theory or principle, but not necessarily in practice. The Developmental Assets framework, from this perspective, can be used as a tool for forging intra- and intersector connections that help build actual communities or community networks. That is, the asset framework introduces common goals and values, supported by a common language, all of which are essential to building community identity and cohesion.

A central function of HCI is to provide a context within which community development can occur within and across sectors. This idea gets back to Larsen's notions of local and metacommunities. A key component of a metacommunity is the cultural glue that binds links across sectors. To the extent that the Developmental Assets framework serves to build such links, it becomes part of a cohesive community bond. It serves to build community within particular sectors, where it might otherwise have been absent, and across sectors where a common language and set of values are necessary for synergistic collaboration.

Leadership Wisdom: HCI's Distributed Leadership Model

Consistent with its focus on local ownership, HCI uses a distributed leadership model. "Community faculty" are HCI-designated experts in community-building efforts. In a staff meeting, Larsen referred to the community faculty as the "tenders" of the initiative, with a critical connection to the continuity and consistency of the community effort over time:

> The crucial assets to any neighborhood are the people that live there ... so that if we're going to have an initiative here, it needs to be owned by people and not by an agency. And so in these local communities what we do is we send somebody in, essentially. In Winter Park it was Sidney, Art Cross in West Orange. And their job is ... to build capacity for an initiative ... And what we've told them to do is look for the people that stick to you ... And then when there's nine or ten of them, bring them into one room and call them the Leadership Committee. And it becomes theirs ... Winter Park found funding to keep the people that were facilitating. The same thing's happened in West Orange. We contracted with Art, for a year, to do that. Now they've found resources to keep Art working with them.

Larsen's comments capture the essence of our Leadership Wisdom thematic cluster. He underscores the importance of local ownership, but links that with a clear bridge to the larger initiative, or what he calls the meta-level community. His view is theoretically grounded, practically implemented, and consistently applied across the larger HCI network. In fact, Orlando's focus on community faculty who could play particularly important local roles in carrying out the larger mission led us to the code *Key Bridges*, which we define as strategically situated HC • HY participants who have the skills and social capital needed to link sectors with each other and to the larger initiative. These are people who have a deep understanding of and commitment to the mission and who have the capacity and will to leverage these traits. Not only are key bridges leaders in and of themselves, but the cultivation of key bridges through an orientation like the community faculty approach in Orlando also represents the wisdom of the initiative's central leadership.

Key Bridges was commonly linked with *Leadership Determination* in our analyses. The latter code refers to a particular quality of many effective leaders found in our study. *Leadership Determination* captures the intensity of drive commonly needed to keep the HC • HY mission moving forward in the face of serious obstacles, such as funding crises, the splintering of sectors due to philosophical divides, or simple everyday fatigue among initiative staff. Key bridges often exercise this type of determination. These are the people who will not accept "no" for an answer, or for whom "good enough" is not adequate. They push the work to the next level through a combination of their status and reputation, along with the intensity of their drive and commitment to the mission.

White Guys Over 50

Larsen and his team did not assume that determined leaders were simply out there for the picking. They recognized the need to cultivate such leaders, such key bridges. At the same time, they knew the importance of shifting away from the usual cast of characters who get slotted for such roles. Larsen recalled:

> Before there was ever a survey done [i.e., the Search Institute *Profiles of Student Life: Attitudes and Behaviors* survey] or the community event that released the survey [results], [a]community group worked for about a year and a half to two years in preparation. We called it community capacity building, and [we] made the links . . . we . . . chose the people that came to their community [meetings] very carefully. There weren't a lot of social workers and teachers [selected], because those are the people that come to all these things. We really worked hard . . . if you'll pardon the expression, it's true for all of central Florida and certainly for Winter Park, we needed white guys over 50. That's who we needed to come to this room if we were going to start having [a real impact on the power brokers living in those communities].

Larsen and the HCI leadership team did not expect these "white guys over 50" to necessarily feel an intuitive connection to the initiative, nor did they overemphasize the importance of this group or any one constituency. Rather, they strongly felt that a new approach would require a new set of players, a more diverse team than one typically finds in community and youth development work. White men with power in their local communities needed to be part of that team, according to the thinking, from the outset. If they were recruited after the fact, they would be more likely to resist, viewing the mission as yet another reform effort of the education and social service sectors. Soliciting their involvement as key bridges to their communities, on the other hand, would help them view the mission as their own, thereby enhancing their determination in carrying it out. One staff member remembered the following exchange:

> When Mike [a member of the HCI leadership team] walked down and said, "You guys didn't do well [recruiting community members] because there were Lexuses and Cadillacs and Jaguars in the parking lot," we said, "That's exactly the crowd we were going after." Because they're often left out when we have, like, community events.

HCI began community-building efforts in Winter Park, a middle- to upper-middle-class area, to demonstrate that the initiative was not targeting low-income communities only. It moved next to West Orange, "on the other side of town." An HCI staff member reported in an individual interview:

> I think it's more intentional [than casting seeds]. Certainly in some sense we're casting seeds . . . we'll do a presentation about assets just about anywhere they ask us to, but we've been very specific about [targeting a range of communities] . . . One of the tips we gave [in

> a recent community presentation] was to go slow . . . What we meant by that was that when we work within one of these subcommunities, an example would be this one named Winter Park . . . we spent a whole year doing what we called building capacity for an initiative there. And what that means is lots of presentations and recognizing and identifying and bringing together leaders for a steering committee for that subcommunity. When that group came together we changed our role; Healthy Communities changed its role to facilitating that homegrown leadership group. They began to plan how they wanted to spin this off and they decided to survey their high school and then to have a community event and mobilize from there. So, we've been very intentional about that and by intentional I mean that we chose our most affluent subcommunity to begin this with because we felt that would help us get past "it's-not-our-kids problems."

On two levels, then, HCI promoted a very different angle on diversity. The initiative diversified efforts by incorporating community affluence and individual economic and political power into its HC • HY model. The message was clear: Not only was this initiative aiming beyond schools and the social service sectors, it was also aiming beyond low-income communities. While poverty contributes extreme stressors that can inhibit healthy development, barriers to such development exist across all communities. The early targeting of Winter Park makes that point vividly in Orlando, just as the targeting of its wealthy residents proclaims that the responsibility for community change lies with everyone.

Youth as Civic Leaders

Keeping kids busy does not mean they feel like they are contributing.

The roles of youth in the local HC • HY initiatives capture an important aspect of Leadership Wisdom, particularly from a distributed leadership perspective such as Orlando's. In all eight initiatives, youth were recognized as playing a unique role in translating the Developmental Assets to language and experience that are "youth friendly." The *Peer to Peer* code, while primarily created to capture the importance of youth collaboration across all the initiatives, can apply to Leadership Wisdom as well: Young people are critical to leading one another, and certain youth exhibit particular acumen in wisely leading their peers and, in many cases, adults. Youth who lead often teach adults how to reach young people effectively by contributing to real-life models of successful adult–youth communication. As adults become more comfortable in their interactions with particular youth, they build the confidence needed to reach out to others; this seems especially relevant for adults who do not work in the education or youth service sectors, including some of HCI's "white guys over 50."

The code *Doing With, Not Providing For* captures the stance that optimal youth involvement requires a reframing of young people from service recipients to co-leaders with adults in processes of community change. This shift is easier to grasp philosophically than it is to put into practice. Across the initiatives in this study, we heard repeated examples of adults providing opportunities or services *for* youth. In many such cases, youth involvement in these adult-provided activities typically was viewed as "active" youth engagement in the initiative. But actively engaging youth to work alongside adults is a different experience, one that requires relinquishing control and exercising patience, according to HCI staff. Because adults are accustomed to being in control, at least hierarchically, in their relationships with young people, it is natural for them to settle disagreements authoritatively rather than collaboratively. In many circumstances, such as parenting or teaching, it may not only feel natural for adults to impose authority, it is also traditionally recommended. Altering such a fundamental life-learned stance takes work. Nonetheless, we found that initiatives such as HCI, which emphasized a "doing with" approach, had more balanced, reciprocal relationships between youth and adults.

The connection between Developmental Assets and HCI's larger mission of community sustainability is perhaps best illustrated by the Legacy Venture Team, a youth philanthropy initiative. The venture team awarded $75,000 in 2002 to youth initiatives focusing on Developmental Assets. The youth team was trained in asset-based grant making, after which it was allowed a great deal of autonomy in all aspects of the team's decision making. The tremendous appeal of this project for youth is that it provides an opportunity to harness youth creativity and knowledge in the day-to-day operationalizing of the Developmental Assets framework.

Consistent with HCI's initiative-within-initiative approach, the youth philanthropy initiative was divided into subteams, or subcommunities, as well. Each of these subteams awarded funds explicitly focused on a particular asset area. The Jewish Community Center's team focus was on diversity, the Boone High School team concentrated on community sustainability, and the Winter Park High School team focused on funding activities that covered a wide range of the 40 assets. The larger Legacy team was trained in assessing different aspects of the grant applications, such as the time line, budget, purpose of the project, how it would be implemented, and its role in promoting asset building. Being able to see the results of their work through the programming they had chosen to support reinforced a sense of ownership for youth within HCI.

The success of the Legacy Venture Team required not only the intelligence and creativity of Orlando's young people but also the capacity of HCI's adults to let go of a good deal of control. In doing so, they claimed, youth involved in the initiative raised their own expectations for one another even higher, and forged a deeper sense of trust across the generations. In our youth focus group, we asked about management of differences of opinion with adults, and about the process of negotiating decision making when such differences arose. Overwhelmingly, youth described feeling listened to and valued by adults within HCI, which, they said, built their confidence in reaching out to people of all generations beyond the formal initiative. As an example of their feelings of empowerment, youth in the focus group described a decision to challenge business owners on what was perceived as an overgeneralized view of youth as shoplifters. Strategies for challenging this perceived stereotype were under way, illustrating the energetic leadership emanating from HCI's youth participation.

Fit of the Model

Developmental Assets as the "Lever"

Healthy Community • Healthy Youth is the Developmental Assets initiative within the larger initiative of HCI, whose mission is healthy human development, within an ecological context that includes economic, social, and environmental aspects. However, the asset framework serves a critical function in engaging and mobilizing a community with diminishing natural resources and an undiversified economic base that has recently been dealt a severe economic blow. The precondition for embarking on a pathway of community change is dissatisfaction with the status quo. In this case, it is a community struggling to sustain itself in the face of extremely high population mobility. The asset framework provides a "lever" for addressing a piece of this problem—the impact of high mobility on healthy youth development. A necessary condition for mobilization is the belief that it is within one's capacity to contribute to the desired change. The Developmental Assets have strong appeal in that they have deconstructed the research on thriving factors to simple, doable behaviors. Orlando's HCI distinguishes the Developmental Assets framework from HCI as a "community organizing tool." However, this distinction does not diminish the importance and success of its function as an extremely "user-friendly" tool. Given HCI's insight that subcommunities are "virtual" in nature, the Developmental Assets begin to build cohesion within subcommunities by providing a common language and mission based in research from a credible source. The trainings are didactic, but in addition to providing accessible information, they serve as opportunities for residents to network and explore opportunities together for potential collaboration. The asset framework is a powerful, common thread woven throughout many different types of organizing systems in greater Orlando, such as the Nap Ford Charter School in Callahan, the Beta Center, the Winter Park Health Foundation, and the Youth Philanthropy initiative. It is the flexible application of the framework that supports the overarching HCI mission of building connections within subcommunities, and identifying local resources that can establish a firmer foundation of local ownership.

The Role of Community Faculty

HCI sees one of its functions as catalytic. It assesses receptivity to the need for change through asset trainings. It identifies strong local leadership and then recedes into a different but critical role of communicating explicit links among systems using the Developmental Assets framework throughout the greater Orlando community. The "community faculty" who conduct the trainings and engage the subcommunities are experienced and skilled leaders, with long histories of activism in the greater Orlando community and a deep commitment to the Developmental

Assets framework. Further, they hold considerable social capital within their communities, which affords them bridging potential to stakeholders needed for the success of the initiative. Community faculty are lawyers, owners of small businesses, university faculty, real estate agents, and former staff of the city planning office. The strong female leadership within HCI had its beginnings in the Junior League, of which several community faculty have been past presidents. Making explicit links among key stakeholders committed to Developmental Assets is one of the ways in which community faculty generates momentum and furthers "buy-in" within the greater Orlando area.

Closing Thoughts

Establishing HCI within an affluent community, Winter Park, was intended to model that community revitalization was not "code" for deteriorating neighborhoods and "at-risk" youth. HCI is also more than Developmental Assets and healthy youth development. The establishment of a Saging Center represents community building efforts with seniors as the target population. The goal of HCI is to be a "partner of integrity" with all the initiatives within the community revitalization movement in central Florida. In this role, it would ideally provide constructive feedback on tough issues and have the respect "of the local resident or the mayor."

Postscript

The Healthy Communities Initiative no longer formally exists, but evidence of the continuing impact of the Developmental Assets is present: The 2009 Mayor's Faith-Based and Community Matching Grants includes the building of Developmental Assets as a key approach for youth programming.

Chapter 7
We Are *Not* a Program! St. Louis Park, Minnesota's Children First Initiative

The light should be in no one place, but in the community itself.
(Focus group participant)

Of the eight community initiatives presented here, St. Louis Park, Minnesota's Children First, has been the most adamant about defending its identity as a "call to action" or even a "way of life," as opposed to seeing itself as a program. This stance has critical implications for the ways in which the initiative works, and it raises important questions for others choosing a similar path. Specifically, because state and related funding often comes with specifications for program development procedures and outcome assessments, Children First has as a rule opted against pursuing project-specific funding. Rather, the initiative leaders view the Children First philosophy and goals as inextricable from the larger community infrastructure. Since the initiative is perceived as a way of life, one challenge to those who seek to learn from its success is to identify precisely how the initiative works—how this way of life is lived. Accordingly, Children First represents a prototype of the Developmental Assets "movement," while simultaneously presenting challenges to identifying and assessing the benefits of such efforts.

We organize our discussion of Children First around two thematic clusters from our larger analysis: the *Faith Factor* and *Paradoxical Tensions*. Our basic code *A Movement, Not a Program* captures the heart of the Faith Factor: It articulates the central mission around which the movements' beliefs and values are organized, and, due to the challenges inherent in maintaining a movement orientation, essentially requires that the people carrying it out become missionaries. As mentioned in the preceding paragraph, Children First participants describe a calling to this work, and they spare no opportunity to pass along the word. A great deal of the faith in and commitment to the mission of community-wide youth development work is exemplified throughout this chapter.

Faith in the Children First and larger Healthy Communities • Healthy Youth (HC • HY) approach is frequently challenged, however, by a host of logistical realities related to the status quo culture of youth development programming and funding for such work. As a result, paradoxes arise regarding the ways in which activities are implemented, funds are raised, and the community is involved. Paradoxical tensions include the relationships among the larger community and key leaders of the

M.J. Nakkula et al., *Building Healthy Communities for Positive Youth Development*, The Search Institute Series on Developmentally Attentive Community and Society 7, DOI 10.1007/978-1-4419-5744-3_7, © Springer Science+Business Media, LLC 2010

initiative: Can a grassroots approach coexist with a powerful executive committee, for example? We bring such questions to light throughout this chapter in an effort to underscore some of the complexities that must be recognized and addressed as HC • HY initiatives strive for growth and sustainability and seek to retain the integrity of their philosophical core.

Redefining the Catalytic Context

Children First has in many respects been Search Institute's flagship initiative, given its role as the initial community effort and as such helping to inspire the national HC • HY movement, followed by its history of success as a local manifestation of that movement. As a mature initiative that had been developing for a longer time than the four presented from the Wave I of our study, Children First reflected a stronger, more deeply engrained cultural identity. The Wave I initiatives were characterized by their efforts to develop a "catalytic context" for community change. The development of that context involved raising awareness about Developmental Assets and educating the community to view itself, on the whole, as the agent of change for the healthy development of its youth. This mobilization process often required the engagement of community sectors, organizations, and leaders who did not typically collaborate or see themselves and their work as directly related to youth development. Further, the Wave I initiatives were often initiating the change process amidst a prevailing perspective of youth as being at risk, rather than as rich resources to be nurtured and mentored in community engagement. For that reason, these HC • HY initiatives appeared as unique and discrete entities, easily identified within the community. Children First was in a different place in its development. Although it had an office and full-time coordinator, the people we interviewed did not see the initiative as a "catalytic context" within the community. Rather, they saw St. Louis Park itself as the catalyst for promoting youth development and Children First as emblematic of that community norm. In the words of one participant:

> It's not a program. And its nomenclature, you know, we're using the word "initiative," but we've gotten [beyond] that. It means something to us, something different that's hard to explain . . . it's just kinda . . . it's community. It's [part of] the history of the community.

Children First presented the challenge and opportunity to study an initiative in a later stage of community change. Search Institute's Community Change Pathway hypothesizes that community initiatives will experience the process of change through five stages: *Receptivity, Awareness, Mobilization, Action,* and *Continuity* (Roehlkepartain, 2001). The Wave I initiatives appeared to be somewhere within the Mobilization and Action stages; they had gone beyond awareness of the Developmental Assets framework, had mobilized certain sectors of the community, and were actively engaged in early stages of the work. In September 2002, when we began our study of Children First, the initiative seemed clearly located in the Continuity stage of change, defined as ensuring that the change in orientation toward the community's role in promoting healthy youth development becomes a way of life (*Living It Out* and *Long Haul*, in our coding scheme). Focus group respondents chafed at the initial questions we asked regarding mobilization challenges, which seemed outdated to them, given the level of their initiative development. They were highly practiced at "telling the story" of their growth, which had become codified through numerous publications, workshop presentations, and reports.

A year later, in the fall of 2003, when we returned to St. Louis Park, Children First was in the process of reevaluating the funding strategy that had served it well for the previous 10 years. Some respondents described the initiative as "at a crossroads," and expressed concern that taking a different direction could compromise its identity. Others believed their initiative's identity was firmly established and that expressions of concern were overstated. Overall, it appeared that Children First *was* at a crossroads of some type, perhaps not a critical crossroads linked with survival but certainly one in which it was carefully assessing its options for moving forward. In this vein, our second visit provided an important window into the ways in which community change is iterative, resulting in established initiatives' revisiting earlier stages of their development and redefining their work in important ways, often in response to emerging concerns and priorities. One of the many contributions Children First makes to our community case study is its illustration of the ways in which an established initiative culture must adapt to changing economic and social conditions.

Distinguishing Features of Children First's Identity

> This was an asset-rich community before this initiative. Three decades ago, I mean, this is not like Minnetonka, Wayzata, Plymouth, Orono, Kansas City—all those places we've all been and seen, where there are asset-rich, or asset initiatives under way . . . This community, it needs to be studied deeply. The city . . . the community ed initiative for 20 years in this community, it just is way ahead of other communities. (Focus group participant)

Respondents portrayed St. Louis Park as an activist community committed to a grassroots, communal ethos. Although there are clearly key leaders without whose support the initiative could not have progressed, respondents expressed a striking discomfort in the attribution of leadership. Similarly, there are funders and private donors who contributed to the start-up of Children First, and yet there was a clearly stated de-emphasis on funding. The initiative takes great pride in reporting that the operating costs are "only the coordinator's salary of $60,000" (as of 2003):

> Children First is labor intensive too, but most of the labor is free. Consequently, fund-raising is not a top-ranking concern for Children First advocates. The initiative was never intended to be an expensive operation . . . In keeping with the initiative's underlying philosophy, building trust is more important than building a trust fund; and relationships are more important than revenues. (Ramsey, 1999, p. 35, 36)

Although there are clearly identifiable community programs that are viewed as generated by and with Children First, asking about specific programs sparked a chain reaction of negative associations. "Programs" are adversely associated with "outcomes," which in turn are negatively associated with outside regulation. A focus group participant explained that

> St. Louis Park is a place that puts children first. It's not, "St. Louis Park is a place where the Children First initiative [exists]" ... I'm always struck by the fact that there is sort of this underlying ... sense of servant leadership ... If you're going to involve everybody, then you [the initiative] can't own it.

Questions concerning key leaders, funding strategies, and programmatic achievement were experienced by our respondents as antithetical to the initiative's ethos of communal ownership. Fueled by volunteerism and civic investment, Children First "works," they argue, because all residents own it. Our respondents depicted St. Louis Park as a community that has established a quality of life in which strength-based work is seen as normative and reflected through seamless collaboration among all community sectors and constituencies. Leaders of other initiatives in our study described concern about spreading efforts too thin—a theme we coded *Spread Control*. This theme captures the dilemma that, on the one hand, a critical mass of the community needs to be engaged for an initiative to progress, while, on the other hand, such broad-based approaches can result in feeling like the work is spread "a mile wide and an inch deep" (see Chapter 3). Children First participants did not express this concern; they depicted an initiative that reflects a community norm in which the "spread" has already been achieved.

Another distinguishing feature of Children First's identity is its relationship with Search Institute. This relationship led to an expansion of the initiative founder's original goal—to address the needs of children living in substandard housing—to include "all kids" (see Benson, 2006):

> It was driven by Meadowbrook community, a neighborhood ... We used to use the term "at-risk children," you know, with those 10 characteristics of a family that doesn't have certain things. And I know that's how [Carl Holmstrom, founder of Children First] would narrowly define it when he would speak. So it was focused on the fact that ... kids would not get on the buses at that point because the Meadowbrook kids were on the bus, because it was too tense, too violent. And that was kind of the segment that he was talking about, that he saw at his schools, that these were the kids that were not surviving ... [T]hat was a personal thing in his life, was Meadowbrook ... Before we had this concept "Children First," it was these children and "they're struggling." (Focus group participant)

Consistent with establishing a cultural identity, there are guiding stories, metaphors, and parables that all initiatives reference and hand down in their efforts to produce a coherent narrative of what has transpired and to ensure the continuity of the initiative. One cannot provide a profile of Children First without describing "the speech."

In portraying the beginnings of Children First, respondents unanimously credited former superintendent of schools, Dr. Carl Holmstrom, and his speech to the Rotary Club in March 1992, as the catalyst for revisiting the community's commitment to its youth. Consistent with the ambivalence regarding attributed leadership, respondents offered differing emphases in crediting the man versus the community for igniting the initiative. Some descriptions were similar to other initiatives participants' descriptions of their leaders—passionate people who challenged the community to reflect on the status of youth. Other descriptions portrayed Holmstrom's speech as timed during a period of community readiness for change, and viewed his leadership as more of a "tipping point" in that transition:

> I don't think this discussion's about Carl; I think the discussion's about St. Louis Park. Carl inherited it and so did the rest of us ... There is something unique in St. Louis Park in that sense of community, especially when you consider the fact that we're a first-ring suburb, and we're not a community that's a stand-alone. Ah, there is a sense of community. There is a sense of affiliation with this ... place ... that is somewhat unique. And it was a very ripe environment for Children First to take off in.

> I think the genius of what Carl did was he named it [Children First]. He didn't create it. He named it ... and if those two people [Mr. Braun and Mr. Packard] in that audience at that speech had not gotten up, it would have been another speech, long forgotten ... Carl gave many other good speeches before that and probably after that. But somehow this one resonated ... he named something that was already happening in the community. But in naming it, it kind of empowers everybody to recognize ... it's a "Duh." I mean it's so obvious. But somehow when you name what you're doing, and you put it in a framework that's positive, it's a lot easier to get on board something that ... has a good name and a positive slant ... If those two funders hadn't been there the speech would have been forgotten.

The famous speech focused attention on the growing achievement gap between children with supportive homes and those living in distressed families. Like many of the committed leaders we have encountered in this study, Carl Holmstrom was at a mature stage of his professional career in which he chose to champion a cause he was passionate about, risk his accumulated social capital, and pose challenging questions to the community. The fact that the Rotary Club membership was strongly represented by business leaders ultimately proved instrumental to the engagement of this sector in Children First. In this sense, Holmstrom can be seen as a *key bridge* linking education and housing to the business community:

> He was a "tipping point" because he was a white, male superintendent ... I worked with him to get the financial support out of those people. They were white, Republican, conservative, I think, right on down the list. I'm not trying to represent this as good in terms of how the world is. It's just a realistic statement. (Focus group participant)

Following Holmstrom's speech to the St. Louis Park Rotary Club, an "At-Risk Families Committee" was formed that began to meet regularly with Peter Benson, president of Search Institute. These meetings served to transform Holmstrom's original vision of supporting youth in high-risk situations into an orientation of supporting all youth. The meetings also forged Children First's unique relationship with Search Institute in a way that has served both the local initiative and the national HC • HY initiative. Children First has been a model initiative, or "lab" for Search Institute, in shaping its national HC • HY agenda, and Search Institute has provided research expertise and professional credibility that has garnered financial and other support for Children First. According to one participant:

> Search needed us and we needed Search. I mean, that's where Peter came to the table and was right at the podium with Carl ... What Search had learned and knew was simply cognitive—the troubled journey. And what Carl knew was reality [but he] didn't know where to go. And, so, when they would both stand up in front of the groups, that's when it began to take hold in whole form.

In 1992, the start-up grant from the Rotarians was used, in part, to contract with Search Institute to conduct background research and to develop a proposal for a planning grant. The research was conducted to identify the issues of the community, to interview experts, and to review literature and innovative models of healthy community development. The At-Risk Families Committee began to engage a larger community representation of parents and business leaders, and became a community-wide Steering Committee. General Mills and the Rotary Club provided the initial funding to raise awareness and to mobilize the community in what became Children First.

The unique relationship with Search Institute has benefited Children First in a number of ways. The financial support, research labor and expertise, and publicity have facilitated, and perhaps potentiated, the action of a community ready for change. The conditions of a healthy economy, Carl Holmstrom's leadership, and an established legacy of innovation and community activism were aligned to provide Search Institute with a fertile testing ground in which to expand upon the theory of community transformation in support of strength-based youth work. The donations of the Rotarians provided funds and the leadership support that enabled Search Institute to conduct grassroots research in the community. The initiative was able to engage the support of key leaders as well as the critical mass of everyday citizens needed to mobilize a movement. The community's history had set the stage for the *synergistic commitment* of organizations and community sectors we described in Chapter 2 as a central activity for creating and mobilizing a catalytic context for community change.

In describing its structural organization, and the extent to which Children First has succeeded in engaging different community sectors in collaboration, Search Institute is viewed as one important sector. This interpretation need not diminish the unique features that made St. Louis Park an ideal community to embrace innovation on behalf of positive youth development. It must be acknowledged, however, that a relationship with a nationally recognized research institute affords opportunities that have shaped Children First in a way that is different from other HC • HY initiatives. Other communities may take longer to move through the Change Pathway because they do not enjoy this partnership. Further, Children First has been instrumental in shaping the national agenda of Search Institute. It is a unique standard-bearer and pioneer of the HC • HY movement. Nonetheless, there is a striking persistence within Children First to de-emphasize the role that money, prestigious affiliation, and key leadership have played in its history—an aspect of its identity that is deeply reflected in its structural organization, which is simpler and less hierarchical than is the case for most other HC • HY initiatives.

It is somewhat ironic that although Children First downplays fund-raising as a core activity of the initiative, Carl Holmstrom chose the business community as the first audience for expressing his concern about youth, rather than another organization, such as the PTA. The decision proved effective. Children First gained strong support from private donors for its first 10 years. The initiative was seeded through the donations of Rotarians Braun and Packard, who attended "the speech," and their continued support has served to sustain a large part of the operating budget. As a focus participant explained, "I think it's part of the story that . . . we've had the luxury, and . . . it's been somewhat easy for us to turn down . . . outside money, because we have been able to count on that money."

Most HC • HY initiatives within this study and beyond wrestle with reliance on state and federal support. Children First has taken a strong position not to seek funding from outside sources. Using outside sources of funding is seen as inviting a form of "bottomless" dependency, a view that echoes the "golden handcuffs" theme expressed by the McPherson, Kansas, community (see Chapter 9). A St. Louis Park participant said, "We've been tempted several times to take government money . . . we don't take government [funds] because then there's a program."

Regulation from interests outside the community is seen as a threat to Children First's identity as an "initiative," given that state and federal agencies operate by funding programs, and outside agencies assess programmatic impact through measures of "outcomes." Focus group participants shared the following viewpoints:

> It was St. Louis Park, and we stayed centered on St. Louis Park. And if we're taking money from Hennepin County or other places, we're beholden to them in some sort of way, even if they say we are not. And to just keep an arm's length from that . . . was really important for us . . . not becoming a program . . . and being an initiative and being grass roots . . . It would have been easy to take the money, and we probably needed the money . . . but we were adamant that it was going to stay St. Louis Park . . . And so it didn't feel right . . . that we should be looking for grants. We're not a grant-raising community.

> Well, there were some opportunities where various different levels of government wanted us to get pulled into some things and effectively start to tie us to some programs . . . We had vigorous debates about those kinds of things, and said, "No, we have to stay an initiative . . . we can't get tied into becoming a program." And it's very easy that we could have become a program through government funding of one sort or another, ah, at a variety of different occasions. And we've always said, "No, we wouldn't do that."

The economic climate of the first decade of the 21st century has been considerably slower than that of the early 1990s, when Children First began. At the time of our second visit, Children First was anticipating a decrease in private donations and reconsidering its funding strategy. At the same time, a 9th-grade program at St. Louis Park High School, which was externally funded by a large state agency, was gaining national notice owing to its positive outcomes. The juxtaposition of these events has presented an interesting opportunity for the initiative to reconsider its stance on external funding, as well as the sustainability of Children First.

Structural Organization: Vision Team and Executive Committee

Children First prides itself on having an intentionally simple organizational structure that is designed to feature and support the work of its "Vision Team." The Vision Team is emblematic of what respondents described as the grassroots nature of Children First. It is the vehicle for identifying problems, generating ideas, fostering collaborative efforts, and affirming successes in the community in support of asset building. Membership on the Vision Team is open to all residents and youth without requirements for attendance or participation. The "Executive Committee" is viewed as the management team of the initiative. It has a prescribed membership of representatives from the "five partners" or community sectors: city government, schools, businesses, faith, and health care. In addition to the five partners, a representative of Search Institute and the Children First coordinator are also members of the Executive Committee. The role of the Executive Committee is to handle routine procedural, policy, and budgetary matters. Members of the committee also serve on the Vision Team, and in that sense serve as structural and functional bridges. In the opinion of one committee member:

> One of the reasons we've been successful is that we balance. Even though it's grass roots, we balance the grass roots and the power piece. I mean, there is a nice balance. Part of the power piece ... [is] that the Executive Committee keeps the crap out of the way for the Vision Team so that they're not saying, "How should we fund-raise? What should our bylaws be?" We keep ... that out of the way so they can truly vision where we need to go. And I think our success also lies in the fact that we've stayed an initiative [rather than becoming a program].

The coordinator position, the only paid position in Children First, is described as a catalyst who stimulates residents' involvement in asset-promoting opportunities across the community. The qualifications listed include "Visionary" and "Thinks innovatively" (www.childrenfirst.org). The many duties of the coordinator include being the spokesperson for Children First, providing training and technical assistance to asset-building efforts, and seeking opportunities to build connections between Children First and existing organizations. The coordinator also functions as the central interface between the Vision Team and the Executive Committee, the latter of which holds supervisory control of the position.

The Desire to be Invisible

Respondents were clearly uncomfortable with and frustrated by focus group questions premised on the belief that initiatives do not progress without the engagement of influential, key leaders. Attempts to make the role of the Executive Committee explicit or visible through our inquiry were at odds with the cultural belief of Children First that understanding the initiative's work lies in observing the products of its achievement, rather than its leadership:

> If anything feels awkward here, it is, ah, we in this interaction today should not represent more than 2% to 5% of whatever your input or feedback is. I think you've got 95% of your job to do, to go to Meadowbrook and go to the free clinic and to go to the ninth grade class ... That's what Children First is. This is ... this feels way ... I feel very awkward here. I mean, people don't even know and care about the Executive Committee, right? (Focus group member)

The desire to remain "behind the scenes" is intended to support the initiative's core belief that the sustainability of Children First lies in the civic engagement and volunteerism that come with ordinary community residents' *owning* the initiative. An emphasis on key leadership smacked of elitism for many of the initiative's participants, and gets to the heart of two related paradoxical tensions in our analysis. The tension of *Anybody & Everybody* by *Key Bridges* suggests that while the work of the HC • HY initiatives holds valuable roles for everyone in the community, particularly influential people are critical to building key connections across community sectors or constituencies. On a related note, *Everyone's Responsibility* by *Leadership Determination* suggests that while it is important for all members of the community to take responsibility for the contributions they can make, it also is essential that leaders of the initiative, whether formal or informal, exercise determined leadership, especially in the face of heightened challenges. Although those involved with Children First can name poignant examples of key bridges and determined leadership, they shy away from such acknowledgments, preferring instead to focus on the front end of the two paradoxical tensions—*Anybody and Everybody* and *Everyone's Responsibility*—the side of their identity they find easier to embrace.

Sector Connection and Representation

A core requirement of the national HC • HY initiative is that at least three community sectors collaboratively mobilize to build Developmental Assets. *Sector Connection* is a descriptive code that emerged in the Wave I of our study. It captures the specific ways in which community sectors connect with the initiative and with each other (Table 2, Chapter 2). Learning which community sectors had partnered in the work reflected the nature and even the degree to which the initiative efforts had mobilized, and the extent to which these efforts reflected representation of both adults and youth.

The structural organization of Children First has sector representation built in through the Vision Team and Executive Committee, both of which include membership from the five partners: city government, business, education, health care, and the faith community. In addition, the Vision Team recruits adults and youth from across the community. The prototype for the representation and collaborative partnership of community sectors was established through the initiative's 1994 "visioning" project, "Vision St. Louis Park." Several of the HC • HY initiatives described participation in such visioning projects, through which town meetings and focus groups were conducted to assess how residents viewed their local needs and priorities. Some visioning efforts began in response to youth risk-behavior statistics.

Others were viewed as a proactive assessment tool for regularly monitoring community progress. All visioning projects strived to include wide representation from the community:

> We started the Vision. And in fact there were no—we did this on purpose—there were no city council elected representatives on those committees. These were all community people . . . We didn't want . . . what happens . . . when the city council member shows up and is on the committee and everybody just kind of defers to that person. We didn't want that. We wanted to hear from the man on the street. (Participant interview)

> The steering committee—again, a process of folks from the community—identified community connections in neighborhoods as being one of the things we needed to work on . . . So one of the solutions that people came up with—and I thought it was a good one—was to organize neighborhood groups, to organize people around the notion of simply living in the same general area . . . It's easy to galvanize people's support when there is some big-time issue that's happening that people don't like or something's that's going on that people don't like . . . But when the issue's been resolved . . . the thing dissipates [it's over]. And . . . that really wasn't the model we were looking for. (Participant interview)

As noted, Children First was initiated through the early engagement of the Rotarians, who largely represented the community's business sector. From its inception, Children First has modeled the connection of business and education through its operating structure. At the time of our visits, the Vision Team meetings we observed were cochaired by a banker and a high school student. Other activities linking business and education have included a businessman's sharing lunch and driving-range practice with six boys. Additional links involved a bank president volunteering in an elementary school weekly to listen to students read, employees of a real estate company working with youth in a beautification project for a low-income neighborhood, and a hotel providing etiquette training and lunch to high school students. Asset building is also supported by the business sector through adapting work schedules to accommodate student employees. The extent of these business sector connections was unique among the eight initiatives in our study, although the initiative in McPherson, Kansas, also drew heavily on involvement from the business sector (see Chapter 9).

Park Nicollet Health Services, a major health industry with headquarters in St. Louis Park, has been a strong champion of asset building, and the Park Nicollet Foundation has been an organizational leader for Children First. Developmental Assets are incorporated into its strategic plan. The president of the foundation has chaired the Executive Committee, and community groups seeking funding from the foundation must demonstrate the promotion of Developmental Assets in their proposals. In addition to modeling and advocating for Developmental Assets within its own structural organization, Park Nicollet Health Services Center has promoted and supported other strength-based initiatives in St. Louis Park, such as the Central Clinic, which operates two and a half days a week to provide services to children and teens at no charge. Providing health care for families without health insurance or the income to pay for medical services is part of the Meadowbrook Collaborative. The transformation of the Meadowbrook community, which will be described later in more detail, is a stunning collaborative achievement on behalf of strength-based work. Further, this collaboration has served to shed light on the need for better representation of distressed youth and families in asset-building efforts.

Defining Achievements and Challenges of Children First

The many partnerships and achievements of Children First have been well documented (Ramsey, 1999; Roehlkepartain, Benson, & Sesma, 2003). For our purposes here, we feature a particularly significant partnership achievement—the Meadowbrook Collaborative—which serves as an interesting model for engaging segments of the community frequently overlooked in youth development initiatives. In a community like St Louis Park, in which less than 15% of the households have school-aged children (U.S. Bureau of the Census, 2000), the capacity to retain a focus on education and youth development is a striking accomplishment:

> We have 12,000 households in St Louis Park, and only about 15% of those have kids in school. And yet when you float a bond referendum the votes have been overwhelming ... We have a lot of senior citizens in St. Louis Park that are involved in community ed programs and programs that have a direct relationship to the school. I mean, the notion of "lifelong learning" in St. Louis Park is a big deal. It's a big, big deal. (Participant interview)

Respondents repeatedly mentioned two factors they viewed as critical to maintaining the community's commitment to education. One is the sociogeographic reality that St. Louis Park shares a city boundary with the north end of Minneapolis, with its fairly large and complex public school system. The other is the strong Jewish constituency in the community, and its historical emphasis on education. A focus group participant shared the following perspective:

> The people that were coming here [new arrivals to St. Louis Park] were not the most elite people. It was more of a service, blue-collar kind of folks that put their value systems out front and were looking for the schools. Simultaneously, there was also a fairly significant dislocation of the Jewish community in north Minneapolis to St Louis Park, and they were coming for very much the same reasons. And they brought with them very, very high expectations about education.

The Meadowbrook Collaborative and Neighborhood Restoration

Built on marshland in 1949, Meadowbrook Manor was the first apartment complex in St. Louis Park and the largest in the Midwest. While the city was a haven for Jews who, because of anti-Semitism, had been refused mortgages in Minneapolis, St. Louis Park historically had not been welcoming to people who could not afford a particular type of dwelling. According to a focus group member:

> One of the most heated controversies of the [past] decade was the question of apartment buildings. There was a market, and developers were eager to build them, but the residents—old and new—staunchly resisted their intrusion into what they felt should be a suburb of single-family homes.

One Children First participant summarized long-held attitudes of some landlords in St. Louis Park by citing a published account from the city's bicentennial commission:

> People renting today are [viewed] as a different type. Landlords are not renting to families with children because they [believe parents] are not raising their children properly. [Their attitude is] "we don't want that kind of people." (Posely, 1976, p. 44)

In 1995, Vision St. Louis Park responded to this attitude by prioritizing the need for more affordable housing in the community:

> Any residential community can only be as good as its housing stock. For this reason, we addressed issues that are critical to maintaining quality housing in St. Louis Park: how to provide a fair share of low-income housing while enhancing neighborhoods, how to create more "move-up" housing, how to ensure safe and well-maintained homes, and how to support quality multi-family housing. (Vision St. Louis Park, 1995, p. 9)

Since neighborhoods are commonly viewed as the essential unit for fostering connectedness among people as a community, St. Louis Park has chosen to refurbish blighted housing whenever possible, rather than removing it:

> At one point there was real serious discussion because these houses were really run down. It was single-family houses, but a lot of rental ... But this was really low-income and some of these houses were really in rough shape. And there was serious discussion about just having the city do a huge bond and buy the whole 100 houses, buy them all, flatten them, and put in new multi-unit housing ... And the neighborhood said, "We'd really like you to not do that. We would like to have you reinvest in our ... we'd like to really try to fix these houses up. We think we have a good community here." And actually it was about the same time as Vision St. Louis Park was going on. And we found some money at the legislature. And in point of fact, if you drive through there now, the places are getting fixed up. (Participant interview)

By the late 1980s and early 1990s, Meadowbrook had become a low-income housing community of 556 rental units and 1,700 residents, and was isolated from the larger community by heavily traveled streets and an industrial property. Dilapidated conditions, slum lord ownership, transient tenants, and a lack of constructive activities for youth made it fertile ground for drug trafficking, theft, and assaults:

> The squad car would pull into Meadowbrook ... and people would scuttle for the corners like rats off a sinking ship. And there would be a guy laying in the middle of the street bleeding ... It happened at three o'clock in the afternoon in the middle of the village square and "nobody saw nothing." People were terrified. Terrified! (Participant interview)

But concerted and consistent actions eventually turned the situation around. Meadowbrook Collaborative is one of the crowning achievements of St. Louis Park. As described earlier, it was concern for the youth of this community that led Carl Holmstrom to make his speech before the St. Louis Park Rotary Club:

> Meadowbrook Collaborative was kind of growing up at about the same time we started Vision St. Louis Park. But Meadowbrook was first. [It began] because we knew that there was a problem. And Meadowbrook has changed enormously over the last [5–10 years] ... I mean, Meadowbrook was called "Ghettobrook" ... it was a community that was falling apart ... a place where the police wouldn't even go by themselves. (Participant interview)

The Meadowbrook Collaborative began in 1993 in response to a group of Meadowbrook parents speaking out in a community forum about their living conditions. The most urgent concern was the lack of safety. In response, an outreach coordinator was hired, and a city recreation program was implemented on-site in the housing community. Consistent with the community policing initiatives adopted by many communities, a full-time police presence was introduced into Meadowbrook to improve communication and promote the development of positive relationships between residents and law enforcement officers. A Crime-Free Multi-Housing (CFMH) model program was initiated in Meadowbrook, which provided property management training to halt the revolving door of criminally involved tenants cycling through housing developments. Finally, a security assessment of the community was conducted that resulted in better lighting and safety improvements, such as dead-bolt locks on exterior doors and anti-lift devices for all windows.

The early years of the collaborative focused on improved safety and needed services, including accessible family health care and out-of-school-time programming. Along with these improvements came improved tenant–landlord relationships, which was fundamental to cultivating a caring sense of community. In more recent years, the collaborative has emphasized the empowerment of residents who were involved in the decision to construct a new playground and, later, a community center. An unusual partnership of private funding and city services led to the construction of "a public playground on private property," as one initiative member put it. The new playground and updated basketball court were built with private donations, but the city removed the former court, planted grass seed to provide an open play space, and maintains the playground facilities. The property owner supported the remodeling of two housing units into a community center. Donations of labor, funds, and professional expertise supported the conversion of the units into a positive focal point for the community that includes a library, a computer center, kitchen facilities, and space for indoor activities.

Partnership with St. Louis Park Public Schools led to the Homework Help program, which was staffed by teachers representing every school building in the St. Louis Park school district, according to initiative members. Having school staff come to the community provided a casual context in which teachers and students could develop a relationship that extended beyond the classroom, and modeled a positive connection between school and the home community. Academic help was more accessible to students, and teachers reported feeling that they had gotten to know their Homework Help students better as a result of working with them in their home environment as well as in the classroom. Homework Help also facilitated the transition from elementary to middle school, as students would become familiar with middle school teaching staff. Although direct parent contact was less extensive than desired, teachers working in the immediate neighborhood increased the opportunity for parents' involvement in their children's education, a well-documented factor contributing to students' academic achievement and positive connection to school.

The range of accomplishments associated with the Meadowbrook Collaborative serves to reinforce participants' faith in the mission of a broad-based initiative approach versus more targeted programming. Although programmatic efforts contributed pieces to the puzzle of Meadowbrook, the larger process was one of public–private community action. Sustaining the collaborative's success would require ongoing commitments from this team. The tension, however, is that each part of the puzzle costs something. If fund-raising efforts are undertaken to sustain the separate pieces, it may increase the chances of their survival; at the same time, piecemeal funding strategies risk placing certain programs at the center and inadvertently undermine the larger community collaboration. It is precisely this type of paradoxical tension that forces initiatives like Children First constantly to reassess the relationship between their faith in the mission and the realities of sustainability.

Transience and the Challenge of Diversity for Children First

A challenge for many HC • HY initiatives is engaging a characteristically transient constituency in strength-based work that relies on relational continuity (see Chapter 6). By its very nature, the HC • HY initiative is intended to build and sustain connections among community sectors, organizations, and people in support of healthy youth development. Making and sustaining these connections implies investing in the community. This understanding was indicated in the suburban vision articulated by the "town fathers" in which resistance to renters is rooted in concern over a lack of community investment. In an interview, the St. Louis Park mayor addressed the issue of transience and the need to adapt youth development efforts to the realities of different neighborhoods and constituencies:

> It [youth development work] is harder ... but ... for example, we had the National Night Out and there was a huge turnout. I mean, yes, I think you can do it. It looks different than if you have a neighborhood where people have been in these houses for 20 years, and they're not renting, they own, they've got very, very deep roots ... And if you have people that might only be there on a six-month lease or a yearlong lease, or maybe even month to month ... I mean, they don't have that kind of connection. And for all kinds of reasons, they can't tell you today that they're going to be able to have their kid finish school here at the end of the year. They don't know.

The mayor followed this comment with an example of an everyday organizational issue that affects transient youth and their potential for tapping into community assets:

> You sign up for soccer, I don't know, three months before the game starts. And those people don't know three months from now where they're going to be! They cannot tell you. They cannot predict three months from now what's going to happen to them. They're working very low-paying jobs, the economy is weird, they're thinking they're going to get laid off, they don't know ... The bus shows up to pick up the kids to take them to the soccer game and there are some kids there that weren't signed up because they didn't know when the sign-up happened ... but now they are there and so they want to go and play ... It means you have to have greater flexibility and it means you have to have some systems in place that you can use to say, "Listen, I know the paperwork isn't here, but this kid wants to play soccer. Get him on the bus."

The mayor's argument is that for transient and other distressed families the rules of the game for community engagement need to be modified. Maintaining the status quo keeps many youth and their families disconnected from potentially available assets. This argument addresses a critique of many HC • HY initiatives and more narrowly defined youth programs: They seldom reflect the diversity of a community's youth. Children First struggles to engage youth who are not in leadership roles or involved in school activities, including sports. A focus group participant said:

> I think the trouble is there are 200 or 300 students who are the usual suspects, who are on everything. But there are 5,000 students in our district ... It's the same kid on this [activity], the same kid on the Day One Committee ... And I don't think we do a good job of, ah, getting a nice cross section. Because ... those same 200, 300 kids are the ones whose parents are bringing them.

Since the initiative office is located in the community's mainstream high school, it is not surprising that those students who are active on the Vision Team and in the high school on behalf of Developmental Assets are the "proximal" youth—those who are always near the action, as the Children First coordinator put it. She has been a strong advocate for extending outreach to youth outside the mainstream. On her suggestion, we conducted a focus group with students at an alternative high school, "SHOC" ("Senior High Options at Central"), as part of our study. The SHOC students, who were mostly African American, did not mention Children First and lamented the fact that "there was nothing to do" in St. Louis Park, in much the same way as youth complained in isolated, rural areas of northern Nevada (see Chapter 10). Students described their perception that adults did not like them, and even feared them, unless they were involved in sports, excelled academically, or were student leaders.

The high proportion of older residents in St. Louis Park seems to exacerbate the tension between minority youth and adults. When we spoke with minority youth in other communities involved in our study, we found that they were often disdainful of the elderly, perceiving them as fearful and racist. Students at SHOC shared similar sentiments:

> They always think we're doing something bad ... They're always staring at us like we're doing something ... or ... they'll walk [to the other side of the street], or like they'll get scared of us, and, you know ... half the time we're not doing anything bad ... They think everybody's in gangs' cause you wear a do-rag.

Recruitment of minority youth has been accompanied by the challenge of engaging minority leaders in Children First, and in the larger St. Louis Park more generally, according to Vision Team and Executive Committee members:

> I don't think it's like there's any resistance to it. We don't seem to be able to get our hands around it ... In the school district we don't have such a horrible problem recruiting minorities, specifically African Americans. We have a terrible problem with retention.

> There just isn't the support once they get here, because they don't find it a comfortable place to stay for [the] long term.

The recent departure of the black female school superintendent, who had followed another African American woman in the position, affirmed the problem of minority leadership retention, according to some participants in our study. Assuming a leadership position in St. Louis Park would be challenging owing to the community's history of incumbency alone. Respondents attributed the progress of the city to having "the right people" at the helm, people who understand the cultural identity of community action symbolized by Children First. These people have tended to be white and male, such as Carl Holmstrom. The latest superintendent, Barbara Pulliam, was a psychometrician, interested in programmatic evaluation and achievement outcomes, particularly for minority students. Having come from an urban high school, she was also accustomed to seeking external grant funding to serve the needs of the school. Although we do not know whether the former superintendent's leadership approach overtly clashed with that of the larger community, we do know that a programmatic, or outcome-oriented, approach would not have been in alignment with the philosophy and orientation of Children First. One question raised by this potential disparity is whether an "outsider" status in the community and the lack of understanding or embracing of the local culture, rather than race, gender, socioeconomic status, or ethnicity, are the key factors in minority retention issues within Children First.

Revisiting the Crossroads: Initiative or Program?

> I think it's hard to let go of an old idea ... Can you stay a facilitated process, which is what initiatives are? (Longtime initiative participant)

As noted earlier, when we returned to St. Louis Park in September of 2003, Children First appeared to be at a crossroads. No longer able to rely on the private donations that had supported the initiative in the earlier years, participants were actively reconsidering the issue of external funding. A core ethos of Children First is community ownership, and external funding was seen as potentially compromising the visioning of the initiative and its facilitative mission. The success of an innovative program at St. Louis Park High School designed to address student failure in grade 9 contributed to the debate. A school counselor with 12 years' experience at St. Louis Park High School developed the program, which was initially funded by a Center for Substance Abuse Prevention state incentive grant. Although Children First did not initiate the program, it is viewed as an evolution of the strength-based orientation promoted in the school. The program was funded by an external agency, had reported outcomes, and seemed to have successfully maintained a Developmental Asset approach. The blend of the asset framework with prevention and even intervention approaches raised new questions:

> The asset framework is such a flexible framework, which is very helpful. But in terms of being really useful to high schools, there needs to be more parameters over, you know, like the teachers. Some things need to be prescribed ... "You're going to have to meet an hour each week with each other to be able to identify kids." Where there are some things that can't just be "make it up on your own." Some of the criticisms of Search [fit]: "We bought

> the trainings, we bought the books; we're not seeing the numbers like you're seeing." ... We know the framework, we use the framework, but we don't do just general asset building, and you can't just tell kids "hi" in the hallway [and call that an asset]. (Participant interview)

The 9th-Grade Program is a particularly interesting strength-based example because its goals include the prevention outcomes of reducing alcohol and drug use, truancy, academic failure, and discipline referrals. The program is notable for its proactive orientation toward incoming first-year students' needs and challenges, rather than being a reactive intervention model that is traditionally triggered by emotional or behavioral problems. During the first two weeks of school, teachers and staff meet with incoming students previously identified by junior high staff as in need of support, as well as having all 9th graders complete a questionnaire indicating their needs. Connections are made with all parents early in the year to establish a foundation for a positive relationship. Parent orientation includes topics such as adolescent development and training in the Developmental Assets, rather than the traditional, logistical information of standardized testing schedules and school attendance policy. The 9th-grade class is divided into blocks of 80 students who work with the same core cluster of three teachers. This block system reduces class size and ensures that all students will have frequent contact with their core teachers. "I Time" is a required part of the curriculum that is designed to foster development of a positive school climate. It is a weekly period of 30 minutes devoted to team building, in which students and staff share activities designed to enhance communication and social skills.

Gaining the buy-in of teachers to the 9th-Grade Program was initially a challenge, according to the program founder. Under considerable pressure to improve student achievement outcomes, teachers viewed nonspecific use of Developmental Assets as demeaning of their skills and experience. One teacher made the following observation:

> "Building relationships" is a kind of given, but kind of condescending, you know, where it's like, "Well, yeah, I'm going to be nice to kids, of course I'm going to be nice to kids—*and* I want them to do well in science" ... [Initially] there was not big buy-in on the asset framework.

The program founder offered a similar assessment:

> Oh, it's absolutely an affront ... [I]f it's something looser and just kind of, you know, asset building generally, very broadly defined ... it's like, "That is so insulting that you think I don't know that!"

"I Time" was viewed as a "throwback" to homeroom, which teachers viewed as a waste of time and successfully fought to eliminate. Again from the program founder:

> This wasn't like a community that's like, "Yea! Let's asset-build!" You know? [But] the fact that there were very specific goals—"We're going to reduce academic failure, we're going to reduce truancy, we're going to deal with discipline, we're going to reduce chemical use"—gave it more kind of grit, where they were like, "Okay. Now this makes some sense. This is what we're going to do here."

"I Time" was the one nonnegotiable aspect of the new program. The program founder (and director) was able to gain the acceptance of "I Time" through assuring teachers that their teaching style would not be prescribed. When the program began to reflect positive outcomes, including the student perception that "I Time" was "not stupid," the teachers began to get on board. Until then, however, it was essentially seen as condescending fluff.

Early positive outcomes of the 9th-Grade Program included students showing an increased commitment to school through improved attendance and a reduced academic failure rate. Although risk behaviors continued to increase, the rate of increase slowed (Roehlkepartain et al., 2003). These "outcomes" were clearly "measurable" and seemed linked to the strength-based approach of "the program," raising new levels of debate regarding programs, outcomes, and evaluation within the larger initiative. By virtue of being a successful program within an initiative, the 9th-Grade Program presented an opportunity to revisit some of the polarities experienced by Children First and other HC • HY initiatives. Three of the key tensions can be summarized as initiative versus program, strength based versus deficit based, and flexibility versus prescription.

The 9th-Grade Program, like the Meadowbrook Collaborative, can be viewed as a mini-initiative nested within the larger HC • HY initiative. Each mini-initiative moves the work of Children First forward. The Meadowbrook Collaborative took the lead in addressing the underrepresentation of distressed and minority youth and families in the larger initiative. The 9th-Grade Program poses the possibility of making space for programs as part of the larger asset-building work. Although the 9th-Grade Program uses a specific, somewhat prescribed application of particular developmental assets, it suggests that prescription or structure need not imply a "cookie-cutter" approach.

Similar to the other initiatives in this study, the challenges facing Children First include a national education policy focused on the achievement of high academic standards. *Standards* is a key word here. The adoption of the 9th-Grade Program was designed in part to help teachers and students meet benchmarks associated with national and state mandates, while doing so in a manner that is consistent with a strength-based orientation. To remain viable, according to respondents across all eight communities, local HC • HY initiatives must be flexible enough to meet the needs of given sectors, such as education, while remaining true to their mission:

> I'm not talking about not continuing to do what we've been doing, [implying that we're not finding the initiative effective]. But asking ourselves the question: "How far has it gotten us? What else can it do to bring value? And then what else of value has it generated that we can include in and make a part of what we're doing as we go forward?" (Participant interview)

Implications

Because Children First has been an effective HC • HY community initiative for a substantial period of time, both its accomplishments and challenges have a great deal to teach. By organizing our presentation around the thematic clusters Faith Factor and Paradoxical Tensions, we attempted to highlight the interrelationship of several critical issues that emerged in our study, and most poignantly in Children First. St. Louis Park's initiative fought hard to remain true to its faith in a community-wide initiative versus specific, highly prescribed, outcome-based programs targeting specific groups of "at-risk" youth. Yet, as with other initiatives, the pressure to meet certain demands, such as stronger school performance for a broader range of students, required some questioning of that faith or at least redefining of it. In this regard, Children First foreshadows the theme of blended models for youth development work, which will be featured in subsequent chapters. A blended-model question confronting the larger HC • HY initiative and Developmental Assets movement is this: How can communities integrate the asset framework into prevention and even intervention approaches while remaining loyal to a strength-based orientation? That question will be taken up in some detail in chapter 9–11.

There are many paradoxical tensions manifested in Children First. A strong emphasis on community ownership of the initiative intersects with the existence and obvious power of an executive committee that carries a good deal of the responsibility for getting things done, including securing funding. This is not to imply that there is hypocrisy at work here, but rather to point out the coexistence of these very real tensions. One can certainly endorse an orientation toward community-wide ownership and still have in place a powerful executive committee. We see Children First holding this tension quite well. But they do not hold it comfortably when the reality is pointed out. There is a strong desire to be, and be seen as, a grassroots movement; the Executive Committee's existence and role complicate that desire.

Finally, there is the issue of money. By de-emphasizing programs as part of its larger identity, Children First risks the loss of regional, state, and national funding. Relative to most other HC • HY initiatives, however, it is a risk they have been able to manage. Their work began with the recruitment of the business sector, and many private donations have come from that part of the community. So, while on the one hand Children First has eschewed the task of fund-raising to pay for programmatic work, it has had the luxury of benefiting from generous and committed donors. A moral of this story, it seems, is that the money has to come from somewhere. When it can be provided by private persons, there may be less need to raise program-based funding; when such private funding is unavailable or limited, however, the need to reach out externally would seem to be stronger. We see further examples of this tension in the chapters that follow.

Postscript

Children First continues to focus on engaging adults from all walks of life and all sectors of the community in taking action to build assets with and for kids. The city has been awarded the America's Promise Alliance "100 Best Communities for Young People" award for three successive years.

In 1996, the initiative developed a new model for driving asset building deeper into the community. The Vision Team ceased, and through a series of meetings with community members and leaders, the Asset Champions Network emerged. The basic premise is that individuals, representing their organizations (faith communities, Rotary Club, businesses, police department, city offices, schools, student clubs, neighborhoods, and so on), all make a formal commitment to go through a three-hour training and sign on to be Asset Champions—looking for ways to weave asset building and information about the Developmental Assets into activities and decision-making processes in their organizations. More than 125 individuals have gone through training and are working as Asset Champions. The group gathers quarterly for structured meetings, and participants can join monthly coffee hours at local restaurants or coffeehouses to share what they are doing and to network with each other. A 3-year grant from the Bremer Foundation helped underwrite the expenses associated with shifting to this new model.

In 2008, a group of students, brainstorming with Karen Atkinson, came up with the idea that asset builders are "superheroes for kids." From that remark, a connection was made to a local caricaturist, Mike Christiansen, and a call went out inviting people to nominate local "Superheroes for Kids." Out of the more than 100 nominees, 18 were selected, and caricatures of them were made into a superheroes comic book that was distributed at the annual Children First Ice Cream Social and throughout the schools and other venues. All 18 made guest appearances at the Ice Cream Social, complete with red capes.

In 2009, three Girl Scouts, working on their silver badges, developed an art project. Organizations throughout the community were invited to paint a picture representing one of the 40 Developmental Assets. Artists ranged from a doctor at the local free clinic to a bus driver to a father and daughter who worked on the project together. The 40 paintings were first unveiled at the 2009 Ice Cream Social and are moving to various venues around St Louis Park. A poster of the 40 paintings was made and distributed.

Children First continues to look for new ways to capture the attention of St. Louis Park residents and engage them in asset-building actions.

Chapter 8
Partnering with Prevention: The Lawton/Fort Sill, Oklahoma, Community Coalition

The Lawton/Fort Sill Community Coalition (LFSCC) illustrates the successful merging of prevention and strength-based approaches to healthy youth development—orientations that are often portrayed and sometimes presented as contradictory. When prevention training focuses explicitly on educating youth about risks associated with substance use and abuse and unsafe sexuality, for example, the approach may lack a strength-based orientation. LFSCC works intentionally to integrate problem-based prevention with strength-based development. It does so through a unique partnership between two separate but interrelated communities: the city of Lawton and the neighboring Fort Sill army base. Along with the economic benefits it provides for the area, Fort Sill presents substantial risks for local youth, in response to which LFSCC has built an impressive track record of pragmatic and innovative youth programs and activities. The main activities are prevention oriented, as the Lawton area confronts a host of challenges it associates with the presence of a local illicit sex industry, which is fueled by the proximity of the military base. The most pressing challenges include high levels of teenage pregnancy, early marriage and parenthood, domestic violence, and teenage alcohol, tobacco, and other drug use. Much of LFSCC's prevention programming addresses these issues through strength-based activities, relying heavily on the training and support of youth as peer leaders in the community.

The shared history between Lawton and Fort Sill has made for a natural extension of the Healthy Communities • Healthy Youth (HC • HY) initiative into both communities as they partner to serve youth. In sharp contrast to St. Louis Park, Minnesota's Children First initiative, LFSCC has several paid staff and supports its operations through a wide array of state and federal funding. When asked to illustrate what makes the coalition unique, participants described the many notable threads that are woven together as part of the fabric of the LFSCC initiative. The coalition has been driven by a strong emphasis on diversity and multiculturalism, strong community and sector connectedness, committed and experienced leadership, and a vision of healthier children and families that is supported by LFSCC's many programs. Whereas Children First placed its faith in being a community movement rather than a composite of discrete programs, LFSCC embraces its identity as an integrated network of such programs.

M.J. Nakkula et al., *Building Healthy Communities for Positive Youth Development*, The Search Institute Series on Developmentally Attentive Community and Society 7, DOI 10.1007/978-1-4419-5744-3_8, © Springer Science+Business Media, LLC 2010

The Lawton/Fort Sill Community and Comanche County

The town of Lawton was founded on August 6, 1901, when the Kiowa-Comanche reservation was opened for White settlement. At that time, a lottery was used to distribute the land. Any person wanting a claim for up to 160 acres simply registered for a drawing. The town site was selected by federal authorities and located on a section of prairie south of Fort Sill, which had been in existence since the 1860s. Today, Lawton is the third largest city in Oklahoma, with a population of approximately 93,000.

Fort Sill was established on January 8, 1869, by Major General Philip Sheridan, and construction of the camp began soon after with troops from the 10th Cavalry, a distinguished unit of Black "Buffalo Soldiers," who constructed many of the distinctive stone buildings in that area. In 1911, the School of Fire for the Field Artillery was founded at Fort Sill and continues to operate today as the world-renowned U.S. Army Field Artillery School. At various times, Fort Sill has served as home to the Infantry School of Musketry, the School for Aerial Observers, the Air Service Flying School, and the Army Aviation School. Today as the U.S. Army Field Artillery Center, Fort Sill remains the only active army installation of all the forts on the South Plains built during the Indian wars. In 2003, troops assigned to Fort Sill numbered approximately 18,830 with civilian employees numbering about 7,500.

Lawton and Fort Sill are located in Comanche County in southwestern Oklahoma, approximately 90 miles southwest of Oklahoma City. Lawton/Fort Sill flows southward into a flat plain with the Wichita Mountains serving as a backdrop to the north. Although we considered Lawton a rural community compared to the larger cities in our study, Lawtonians do not necessarily see themselves that way. One focus group participant described the community as follows: "Lawton isn't really a rural [area]. People don't know what to call us. They say, you're tribal, you're rural, you're urban ... it's that weird sort of dichotomy ... It's ... 'like a little town with a big-city feel.' "

In 2000, the total population for the geographic area of Comanche County was 114,996. The county population is approximately 65% White, 19% Black or African American, 9% Hispanic or Latino, 5% American Indian and Native Alaskan, 2% Asian, and less than 1% Native Hawaiian and Other Pacific Islander. The top five economic industries are education, health, and social services; retail trade; manufacturing; arts, entertainment, recreation, accommodation, and food services; and public administration at 24.2, 13.1, 9.8, 9.1, and 8.5%, respectively. Major Lawton area employers include Fort Sill, Goodyear Tire & Rubber Company, Lawton Public Schools, Comanche County Hospital, and Walmart Supercenter. According to LFSCC staff, the local economy reflects the concerns of the larger state: "Oklahoma is a very depressed economy. It's stagnant. Texas calls us a third world country" (Lawton staff focus group). It is precisely this larger picture of economic stagnation that makes Fort Sill so important to the area.

Coalition Features

When we selected LFSCC as a Wave II site for the study, we viewed it as a relatively new initiative. Despite the coalition's recent registration as a formal HC • HY initiative, however, it had a 15-year history of prior work. From the onset, its efforts were rooted in the mission of supporting healthier children and families through programs and strategies that empowered the community to resist high-risk and unhealthy choices. LFSCC began in 1988 as the Lawton Public School's Substance Abuse Committee. Shortly thereafter it became the Citizens' Task Force for a Drug Free Community, and soon after that the Mayor's Commission for a Responsible Drug Free Community. As one community focus group participant put it, "We've been through a couple of name changes since then, trying to diversify." The multiple iterations or brands of the coalition are illustrative parts of its history, yet the emphasis on and mission of prevention for problems such as substance use, violence, teen pregnancy, and tobacco use has remained consistent over time. In the words of the coordinator:

> [The coalition's] mission to date has been primary prevention: to work with schools, parents, peer groups, and the larger community to reduce risk behaviors and promote positive behaviors—assets, protective factors—and also to network with state agencies, and ... on the national and regional scene to bring as many high-quality research-based prevention activities into this part of the state [as possible].

There was a widely held belief among LFSCC participants that all of Lawton's youth are at risk to varying degrees, based on the risk factors spawned by the local environment. This perspective differs dramatically from the philosophy of the Children First initiative presented in Chapter 7, which organizes its work around a view of youth as filled with promise and in need of opportunities and supportive relationships for manifesting that promise. With its historical emphasis on prevention programs, which traditionally are deficit based, it is interesting to consider when and why LFSCC adopted Search Institute's Developmental Assets framework as part of its mission. In the early 1990s, the coalition's executive director began using Developmental Assets materials in talks and programming, viewing the strength-based approach as consistent with a primary prevention philosophy. As will become clearer in this chapter, the integration of prevention and Developmental Assets approaches was energizing for the LFSCC staff, and created new opportunities for bringing additional funding for youth development resources into the Lawton/Fort Sill community.

Strategic Funding

Each initiative approaches fund-raising differently, depending largely on the nature of its work and its organizational structure. Similar to Nevada's Healthy

Communities Coalition of Lyon and Storey Counties (Chapter 10), prevention funding has provided the bulk of the financial support for LFSCC. A federal grant from the Office of Juvenile Justice and Delinquency Prevention and a grant from a Drug Free Community Support assist the coalition's Life Skills Curriculum and Drug Free Fair, along with numerous other programs and activities. A Department of Mental Health and Substance Abuse Services Area Prevention Resource Center grant sustains a marijuana awareness campaign, a tobacco-free coalition, and a violence prevention peer mediation program. Each of these grants supports prevention activities of one type or another.

LFSCC's grant-funded financing strategy differs markedly from the philosophy and approach of Children First, which, as we described in the preceding chapter, emphasizes private donations and community ownership rather than external grant funding. Differences in funding strategies are dictated by community resources as much as they are by initiative philosophy. The depiction of the Lawton area as economically depressed helps explain LFSCC's emphasis on external grant funding. In fact, coalition staff stated that small, local foundations tended to focus on very basic needs of organizations such as space renovations; they choose to "give money mainly to bricks and mortar ... They just haven't been looking to fund any innovative prevention programming."

The absence of large foundations in Comanche County was attributed to the lack of money for buying oil within the immediate community. In parts of Oklahoma not far from Lawton/Fort Sill, large foundations have developed from Scully Oil and Halliburton. According to the coalition coordinator, "We have hardly any local funding at all, which has been—it's probably *the* weakness of the organization, that we don't have a base of local funding." Because there are no large community foundations that specifically fund youth development initiatives, grants received typically come from outside the area, are for prevention work, and, as described in a community feedback conversation on our study, are "geared toward compartmentalized groups of youth." This funding scenario translates into targeting youth who are most vulnerable to engaging in various high-risk behaviors. As such, LFSCC has had to be particularly mindful of strategies for reaching a broader range of young people. A participant attending our study's community feedback conversation summarized the reality succinctly: "We have to get the money we get and leverage it to do lots of different things with lots of different kids." In other words, an organization has to know what funding opportunities best match its mission if it hopes to have a high probability of obtaining that funding. The director explained:

> We just don't write for everything ... I mean, we'll look at things where there is a good chance. Because there's some things ... like not getting this Teen Pregnancy money ... all the people that got it ... are associated with the county health department. If I'd known that I probably wouldn't have even written for it.

LFSCC boasts of an 80% funding rate for its grant proposals, a level of success it attributes to studying the alignment between its mission and the priorities of the foundations, organizations, and individuals distributing the funds. Timing, experience, and confidence contribute to the strategic funding approaches used by the coalition. Timing and awareness of organizational readiness are critical, according to participants in a staff focus group:

> We ... wrote one grant. It was called "Community Empowered Schools" ... and it was based on the Search Institute Developmental Assets model as the logic model ... And we listed the assets we were going to develop, and we listed how we would measure those ... outcomes—short term, long term ... We didn't get funded. And we think we're ahead of the curve ... And then 2 years later the state was saying that, you know, the asset model was what they were going to start doing. But I think we hit them with it before they knew about it ... And so we kind of [backed] off from that as our model when we do our funding, and we go to the protective factors, which is the language that they use ... but our practical model back here at the office is the asset model.

The fund-raising capacity of LFSCC's leadership team is a result of broad-based experiences gleaned from other professional arenas. Collective expertise in grant writing, program management, and a range of youth service work contributes to the coalition's successful strategies for raising money. One staff member, for example, acquired experience in grant writing and youth work through the Illinois Caucus for Adolescents, where she worked with teen mothers and advocated for comprehensive health programs for adolescents, including young parents and their families. Another key staff member got her beginning in grant writing at an Area Prevention Resource Center. New staff members bring such experiences to the coalition and are supported in sharpening their fund-raising and program development skills through the work. In other words, LFSCC sees itself as a highly professional prevention and youth development organization, and in this sense it differs from the more grassroots initiatives in our study.

One of the key LFSCC leaders with responsibilities for training and supporting newer staff members stated the issue quite simply: "My goal for employees here, prevention staff, is that by the time they've worked here for 6 years they can write a grant [from] start to finish." This unabashed orientation toward professional grant-writing competence appears central to the coalition's success. As funding is raised for new programs, new proposals will need to be written to keep things going. LFSCC has taken that reality to heart and is training staff accordingly.

Organizational Structure: A Nested Prevention Network

After a decade of work on specific prevention programs, LFSCC became part of the larger Wichita Mountains Prevention Network (WMPN), the nonprofit agency that has housed and sponsored the coalition since 1999. WMPN works with other state agencies to identify and implement research-based prevention approaches for schools, parents, peer groups, and the larger community. The network grew out of the Area Prevention Resource Center (APRC), which assists communities across Oklahoma in developing effective substance abuse, violence, teen pregnancy, and HIV/AIDS prevention programs for youth. APRC was originally part of the Comanche County Memorial Hospital, where it had been soliciting and receiving Department of Mental Health prevention grants since 1989. In 1999, the APRC broke off from the hospital to form the autonomous nonprofit organization that became the Wichita Mountains Prevention Network.

At approximately the same time, LFSCC was seeking to apply for funding through a Drug Free Community Support grant and needed a nonprofit agency to sponsor it. This pragmatic need spawned the merger of the two organizations, with LFSCC becoming part of WMPN in 1999. When LFSCC came under the wing of WMPN, it brought along the monies it received from the Drug Free Community Support grant. WMPN has its own board of directors and LFSCC has its own leadership and executive committee in addition to a chair, cochair, secretary, treasurer, media coordinator, volunteers, and a number of standing committees. In summary, the work, the funding, and the organizational structure of LFSCC are thoroughly immersed in the world of prevention, making this initiative ideally situated to explore the blending of prevention and strength-based approaches.

Characterizing Themes

Representation and Shaking Up the Status Quo

Many of the issues presented earlier, as well as those that follow, can be viewed as manifestations of two thematic clusters from our analysis presented in Chapter 3. The first cluster, *Representation*, captures the demographic and political involvement of the HC • HY initiatives. We also featured this cluster in our presentations of initiatives from northern Michigan (Chapter 3) and Moorhead, Minnesota (Chapter 4), but the flavor of the cluster is quite different for LFSCC. The focus here is on the racial diversity of the leadership in the initiative, and the extensive involvement of various community sectors in the work. In addition, the focus of a prevention orientation blended with a strength-based approach raises particular issues of professional youth programming representation, which makes LFSCC look very different from many of the other initiatives in our study.

The role of prevention practitioners or professionals in LFSCC brings us to the second thematic cluster. *Shaking Up the Status Quo* originated in our analysis as a means of capturing how the HC • HY initiatives function in ways that depart from typical youth service and programming for youth. However, growing beholden to a pure notion of strength-based approaches can create a new status quo, a new norm, as our larger analysis suggests. LFSCC is emblematic of a potential next wave of community-wide approaches to youth development—approaches that integrate strength-based work with prevention and even intervention efforts. We refer to these efforts as blended models, and view LFSCC as an exemplar of this emergent genre.

The Role and History of Diversity in LFSCC

The Lawton/Fort Sill Community Coalition is the only HC • HY initiative within our case study sample that has an African American in a leadership position. Women make up the majority of staff for these initiatives, but beyond gender, LFSCC is the most diverse. A diversity exhibit, A Struggle for Equality: A Community View, at the Museum of the Great Plains located in Lawton, provided a detailed understanding of the role Fort Sill played in the integration of Lawton and the establishment of African American leadership in the community. African Americans had long-established thriving communities within Lawton, with the churches and schools serving as centers of strong cohesive bonds. In fact, with the advent of busing and school integration in the 1960s, there was strong resistance from the Black community to having its youth dispersed among the White schools.

Fort Sill played a critical role in supporting integration by using economic sanctions (boycotting) against businesses and employers who refused to hire or serve African Americans. The proximity of Fort Sill allowed the Lawton school system to receive large amounts of federal funding called "impact aid," and there was significant pressure to ensure that children of military personnel were not receiving a compromised education. The African American population in Lawton grew to approximately 16% during the early 1970s as a result of the expansion of military and job opportunities for Blacks. This influx affected the housing industry, and antidiscrimination laws were implemented in the sale and rental of housing in Lawton. The urban renewal movement resulted in the resettling of African American families, which in turn, affected the school integration process. Fort Sill threatened to make businesses and areas of Lawton "off limits" to soldiers if the city did not comply with equal opportunity practices, such as renting homes to African Americans. Lawton's heavy economic dependence on the military made this an effective incentive for confronting racial discrimination and moving toward community integration.

Having a visible and accessible pathway to higher education located within the community was another factor that may have contributed to relatively early advancement of African Americans to civic and educational leadership roles in Lawton. During the 1970s, Lawton's Cameron University began to grant bachelor's degrees, and the first recipient was an African American minister who was a strong community activist. By the mid-1970s, the university had a higher percentage of African American students than any other university in the state, except for historically Black Langston College. African American faculty also held powerful positions within the business community, such as serving on the Chamber of Commerce.

Given the racial history of the city, it is not surprising that the local HC • HY initiative may have been ahead of the curve nationally with respect to diverse leadership. Consistent with the diversity in the staffing of LFSCC, the initiative's efforts reach a broad range of the youth population with strong recruitment of Black youth leaders—a hallmark of its success. Philosophically, it is particularly important for a prevention-focused coalition such as this one to have a strength-based orientation. The linking of deficit-based approaches with youth of color is precisely the pattern that the national HC • HY initiative was designed to challenge. Lawton/Fort Sill is taking on that challenge in an especially complex manner through its blending of prevention and Developmental Assets.

Sector Diversity

The Lawton/Fort Sill Community Coalition is composed of diverse sector collaborations. This section will describe the relationships, some reciprocal in nature, between sectors involved in the work of the coalition. As with racial diversity in the initiative, sector diversity also is a clear product of the local history—of the community and the coalition. Most of the partnering sectors have a long-standing history of involvement with the initiative and the issues it addresses. When the coalition started there were 8–10 subgroups within which 30 different organizations or sectors were represented. These included education, the military, the religious community, businesses, youth, elders, social services, youth-serving organizations, and tribal groups. Given the focus on networking within LFSCC, it is understandable that the net was cast particularly wide from the very beginning.

Fort Sill: The Military Sector

The city of Lawton and Fort Sill are considered good neighbors. It is common for residents of Lawton to participate in and volunteer for Fort Sill activities and events, and vice versa. Fort Sill's Youth Service's soccer program shares its fields with the Lawton soccer program, and Fort Sill assists with Lawton's annual Drug Free Fair. The army base sponsors the Wings of Eagles program, a 3-day mini-camp for 6th-grade boys and girls designed to develop self-confidence and refusal and anger management skills, which it extends to the Lawton public schools. As a result of heightened security at Fort Sill following the 9/11 terrorist attacks, a reduction in some forms of community involvement ensued, particularly involvement that brought "outsiders" onto the base. Nonetheless, the military base remains a vital partner in LFSCC's efforts.

Fort Sill's Child and Youth Services maintains a connection to all community schools. There are approximately 1,000 children living at Fort Sill and two K–6 schools on the base that are part of the larger Lawton/Fort Sill school system. A school liaison officer works to maintain strong connections between the larger system and the schools on the base. Youth living at Fort Sill have the option to attend one of the schools on the installation or those in the general community. The school liaison officer assists with choosing a particular school or program, as well as advising on scholarships and special needs. The Fort Sill Child and Youth Services coordinator, who is an active member of LFSCC, stated that the schools on and off the base serve to connect the two communities. Overall, the communication and connection between the military installation and the Lawton community are viewed as strong, which, according to the Fort Sill Child and Youth Services coordinator, is not the reality shared in other parts of the country where military installations and the local communities "ignore each other."

Along with all it provides, Fort Sill also brings its share of challenges to the Lawton community. The high rates of teen pregnancy and substance use, the local "sex industry" (the term used by focus group respondents to refer to Lawton's challenges with prostitution and the proliferation of strip clubs and topless bars), and high mobility are some of the issues associated with the military presence. It was the general opinion of the people we interviewed, however, that many of these concerns were already at issue in the Lawton community, and that the presence of the military seemed to exacerbate them. For instance, a coalition member noted that the introduction of new drugs has been attributed to the mobility associated with the military. As new soldiers pass through, they sometimes leave their mark behind, whether it is a drug habit or a fatherless child. But, according to this member, Lawton has its own history of drug use and teen pregnancy, possibly making it ripe for the military influence.

The sex industry perhaps captures the complexity of the Lawton/Fort Sill relationship most vividly. As LFSCC participants pointed out, the military did not bring this industry to Lawton. Rather, it was people from Lawton who created the industry as a way of capitalizing on the military presence. Further, there is a history of prostitution in Lawton that precedes the existence of Fort Sill and contributes to the broader illicit sex industry. Neighboring Oklahoma cities and towns view Lawton, sometimes referred to as Oklahoma's "Sin City," as "the place to go to have a good time." It was not uncommon for fraternity members from the University of Oklahoma in Norman to drive 75 miles to Lawton because of the prostitution and bars. Legend has it that young women from Lawton who attended the University of Oklahoma or Oklahoma State University were not considered "sorority material" because of the promiscuity associated with "girls from Lawton."

Education

The education sector, particularly Lawton Public Schools, bought into the coalition's efforts early on:

> And they've [Lawton Public Schools] been very supportive of this commission from the get-go. They've been a prime mover and shaker ... That's ancient history now, but they were ... the ... movers and shakers that helped to give this organization credibility way back when. (Focus group member)

It might seem commonsensical that the school system would be a primary partner in the initiative, but that is far from the norm within some initiatives, particularly in cities whose school systems are struggling with high-stakes testing. But our impression is that the larger community of Lawton was so attuned to the risks faced by its youth that it was uniquely receptive to prevention efforts that might help keep students in school rather than losing them to pregnancy or substance abuse. The main way in which the schools have partnered is in opening their doors to prevention programming during the school day and after school. Such school buy-in has strengthened LFSCC's hand in pursuing grants, as the schools have been quick to provide logistical support for grant-funded efforts.

All Kids

Recruiting youth to join, participate, and stay actively engaged is a struggle for many initiatives. That has not been the case for LFSCC, however. Particularly impressive is that youth who are typically involved in leadership roles and extracurricular activities (the youth sometimes referred to as "star kids"), as well as those who have been labeled as troubled or at risk (sometimes referred to as "those kids"), are encouraged to be active and take leadership roles in the coalition. Both groups of youth have been successfully recruited and play key roles.

Youth from Treasure Lake Job Corps partner with LFSCC for several of the coalition's yearly community-wide events. Treasure Lake Job Corps, operated by the U.S. Fish and Wildlife Service on the Wichita Mountains Wildlife Refuge in southwestern Oklahoma, is a residential high school and youth training program for 16- to 24-year-olds. Vocational training is provided in a number of trades such as union cement finishing, union carpentry, plumbing, cooking, and business office technology. While at Job Corps students have an opportunity to obtain a high school diploma or a GED, along with their vocational training. Historically, Job Corps has been designated for those youth viewed as at risk academically or otherwise, but participants in the Treasure Lake program do not necessarily fit that profile. Some of the Job Corps youth already had earned a high school diploma and were involved in vocational training, some were on a delayed military plan, and others were students who had failed in traditional high schools and dropped out.

A number of youth from the Job Corps attend LFSCC meetings and are involved in peer-led prevention activities such as the Drug Free Fair. In some cases, according to coalition staff, certain activities would not have been possible without the volunteer efforts of the Job Corps youth. Overall, there is a strong belief that all youth need opportunities to be in leadership and decision-making roles, and that they need guided practice in building the skills to carry out these roles effectively. LFSCC strives to create such experiences for youth through the provision of safe environments in which successes can be experienced and mistakes made without shaming judgments or ridicule. Although LFSCC presents itself as *youth minded* (i.e., as having the capacity to take the perspective of youth), there are some skeptical coalition members who believe that the coalition, the Lawton/Fort Sill community, and other organizations could be doing more to engage marginalized youth. But relative to many of the other initiatives in our case study, it was clear that LFSCC was quite advanced in the authentic inclusion of "all kids" in its leadership efforts.

Elder Involvement

Elders in LFSCC, who prefer to be called "seasoned" members, as opposed to seniors, play an active and critical role in the coalition. The extent and level of engagement by seasoned members in LFSCC differ from the experience of other initiatives, which have described difficulty securing representation and maintaining the participation of community elders. Two instrumental seasoned members argued that involving older adults in the work requires "tapping into their interests" and continuously communicating with them. Fortunately, a core interest of many elders is spending time with and passing on their wisdom to young people. In a coalition like LFSCC, there is ample opportunity to satisfy that interest.

Two seasoned participants in one of our focus groups discussed seeing many successes and failures through their youth development work within and outside LFSCC, and they stated that "more experienced" participants within the coalition could reduce repetition of the failures if they were matched with newer participants. By providing guidance and support to the "younger generations," according to these participants, repeat mistakes can be avoided and the older members will feel valued and yearn to stay actively involved. As older participants have remained active in LFSCC, they have provided invaluable membership consistency and have come to be the "keepers" of the coalition's and community's history.

City Government

Youth-serving organizations such as the City of Lawton's Youth Services Program, which partners with LFSCC, serve as important connections between the coalition and city government. The program not only provides its own services but also informs youth, parents, and the community about the range of quality programs offered in the Lawton and Fort Sill area. When the Youth Services Program coordinator posted Fort Sill activities on the city of Lawton's youth services Web site, parents would call her asking why activities at Fort Sill were not available to Lawton youth as well. As a result, she began exploring partnership possibilities with Fort Sill's Child and Youth Services coordinator. Together they wrote a memorandum to the Department of the Army requesting approval for sharing Fort Sill youth facilities and activities with civilian youth. Although the reply was still pending at the time of our study, the effort itself captures the spirit of community collaboration and networking that defines LFSCC.

Lagging and Lapsed Representation

In addition to the sectors and constituencies that have been described here, the business and religious communities have played modest roles in the initiative. Fairly substantial financial contributions have come from the business sector over time, but money has largely been this sector's only contribution. LFSCC's leadership would prefer a more active partnership with the business sector, perhaps one that includes apprenticeships and mentoring opportunities for youth. It seems likely that involvement from the business community has been limited owing in part to the professional prevention tone of LFSCC's work. Unlike St. Louis Park, Minnesota's Children First initiative, with its focus on everyday citizens' owning the work, Lawton/Fort Sill's efforts have largely been professionally run, grant-funded activities. A drawback of this approach might be the challenge of involving a greater variety of laypeople.

The faith community was more active in LFSCC originally, but that sector's contributions have been nonexistent in recent years, at least at the larger community level. According to LFSCC staff, many places of worship in Lawton do a great deal of community-building work for their own religious communities, but these are closed activities and as such do not affect a wider youth base. We could not help but wonder whether differences in youth development philosophy contributed to the increased insularity of the religious communities.

Finally, there has been a tribal connection to LFSCC as well, but the details on this connection to the coalition are sparse. The Comanche Tribe has worked with the coalition to a degree through the tribe's substance abuse program, but this seems to have resulted in a simple sharing of information rather than a more active partnership. Despite the lapsed, lagging, or sparse connections to these three sectors, LFSCC has worked hard to engage a broad constituency of participants and sector partners. Their strategic efforts to raise funding for specific prevention efforts have certainly influenced some of their partnership priorities.

Blended Models of Community Change

Three basic codes from the thematic cluster Shaking Up the Status Quo help frame the contribution of LFSCC to our larger study. The code *Change as We Learn* captures the Lawton/Fort Sill coalition's pragmatic approach to meeting the needs of its youth community. The coalition began in a school and moved to several other host organizations prior to settling into the Wichita Mountains Prevention Network. It shifted its home base to better meet its needs, which are the needs of its community. After separate attempts to use prevention and Developmental Assets approaches seemed inadequate, the coalition worked strategically to integrate the two frameworks and as a result has experienced tremendous success in fund-raising and program development efforts.

Funding Strategic and *Impact Assess* are the two other codes from the thematic cluster that apply soundly to LFSCC. The coalition chose community partners that would give it the strongest chance of obtaining issue-specific prevention funding. And the hiring of coalition leadership staff was guided by the perceived ability of prospective leaders to raise money. In the face of serious risks to its teen population, a strategic commitment was made to organize around prevention resources. This commitment comes with the requirement to assess outcomes systematically. LFSCC emphasizes the adoption of research-based prevention strategies, those that have been previously found effective, and then assesses their effectiveness locally. There is not the reticence toward program language and program requirements found in St. Louis Park and some of the other communities in our study. In this sense, LFSCC represents an important application of the HC • HY movement. It is helping raise the question of whether it is better to blend strength-based work with prevention and perhaps even treatment approaches than to take the path of either/or. But what exactly does the blended model look like in practice?

> We teach the 40 Developmental Assets model in the classrooms, all the 7th-grade classrooms that we go into ... It's at the end of the ... abstinence curriculum that we do, and then we do a dating and domestic violence program too [and the assets are included there]. (Staff focus group)

The focus of LFSCC's activities is on proactive programs and strategies that empower the community to develop healthy, responsible, and involved individuals. Its partnerships with local organizations, such as those with the Treasure Lake Job Corps and Fort Sill described earlier, along with countless hours of volunteer time from youth and adults, fuel the programs of the coalition. The coalition director stated the following:

> There are about 100 active volunteers and those are people that are volunteering at least ten hours each ... It's not just drive-by meetings ... we have a lot of adults that are coming. And then the teens ... I know we have at least 50 teen volunteers. But of that 50 there's probably 22 of those kids that every time we open our doors for a meeting are here. I mean, you see certain kids and they're always here for everything.

Many of the coalition's activities fall under the umbrella of prevention advocacy, defined as providing prevention-related information to organizations that can deliver prevention programming. This process includes sharing prevention research findings with schools, parents, and others who might play a role in prevention programming. The advocacy-based networking component brings to light striking parallels between LFSCC's focus on community-wide prevention and Children First's focus on community-wide strength-based development. In both cases, the emphasis is on getting as many constituencies involved as possible.

By framing its prevention activities in Developmental Assets terms, it is easy to see how LFSCC views itself as a legitimate HC • HY initiative despite its thorough prevention focus. The coalition's work is clearly organized around the four external asset categories of Support, Empowerment, Boundaries and Expectations, and Constructive Use of Time. Youth are provided a great deal of support through the committed involvement of the prevention network, they are recruited to take active roles in the work of the network, their peer leadership training focuses very much on holding high behavioral expectations for themselves and others, and their role in the network provides multiple outlets for the constructive use of their time. The following sample of activities captures the integration of prevention and Developmental Assets in the LFSCC model.

The Southwest Tobacco Free Coalition takes a primary prevention approach by educating children and youth on the effects of tobacco use. This educational focus is then complemented by SWAT (Students Working Against Tobacco), a youth-led effort to advocate for antismoking behavior and legislation. Through SWAT, youth of ages 13–18 are supported in uniting to fight against the tobacco industry by organizing to share information in their schools and communities and through letter-writing campaigns targeting political leaders, tobacco companies, the media, and advertisers.

The Reality Check Campaign focuses on primary prevention of marijuana use by educating youth concerning its effects, which includes countering the myth that marijuana is nonaddictive and not harmful to one's health. Youth play an active role in the campaign by helping to share and distribute information. Project Under 21 is an underage drinking campaign with a central focus of reducing alcohol sales to youth. Peer leaders participate in the campaign by helping to share statistics on problems associated with underage drinking through presentations in schools and community groups. The Drug Free Fair is an annual event planned by intergenerational participants for health awareness and the education of families focusing on what parents can do to prevent drug use. Red Ribbon Week is a weeklong awareness campaign to promote abstinence from alcohol, tobacco, and other drugs; as do the other prevention activities listed here, it uses peer leadership as a core strategy. The Life Skills Curriculum is a research-based alcohol, tobacco, and other drug prevention curriculum delivered to 6th-, 7th-, and 8th-grade classes in Lawton's public schools. By focusing on decision-making and peer leadership skills, this prevention approach works from a strength-based or Developmental Assets orientation.

The Postponing Sexual Involvement program recruits teen trainers, ages 15–19, who provide peer education regarding early sexual involvement. Discussions focus on the goals and aspirations of teen girls, and the importance of protecting one's future opportunities by making healthy and thoughtful decisions regarding sexual activity. Girl Power! builds on the theme of girls' strengths and future possibilities through an all-day conference held twice yearly for 5th-grade girls from all of Lawton's public and private schools. The conferences use interactive workshops to provide modeling and education through activities that build self-esteem and motivate students to achieve in school and participate in extracurricular activities, including athletics and career exploration. These proactive developmental messages are linked with prevention training concerning early sexual activity, substance use, and conflict resolution.

Peaceful Resolutions for Oklahoma Students is a research-based peer mediation program focusing on violence prevention for K–12 students. Its goal is to train a broad range of youth as peer leaders in their respective schools and communities. Wings of Eagles is the 3-day mini-camp provided by Fort Sill that was described earlier. Its focus is on supporting 6th-grade boys and girls to develop the self-confidence needed to exercise healthy refusal and anger management skills.

Parent Network is designed to provide training to parents of teens for networking with one another to gain support and for learning how to access community resources. It encourages parental involvement in school events, parenting workshops, and community leadership activities.

Each of the programs briefly described here has a clearly defined prevention focus, and all of them use youth leadership to varying degrees. A drawback to this approach is that a core group of youth leaders experiences the bulk of supportive training relationships with adults, while the neediest youth are essentially recipients of prevention programming. LFSCC is well aware of this risk and views the proactive involvement of a wide range of youth as critical to the continuing evolution of its work. As an example of the coalition's efforts to reach all youth in the community, we close this section of the chapter with an overview of the Rural Abstinence Project.

Results of the Youth Risk Behavior Survey administered to Lawton students in grades 10–12 in 1999 revealed "frightening information" in the area of sexual behavior, according to LFSCC staff. Of students in Lawton high schools, 59% reported engaging in sexual intercourse, and 39% of those sexually active teens reported having four or more partners (the national average is 16%). Oklahoma is 12th in the nation for teenage pregnancies, with Lawton's pregnancy rate 40% above the state average. Even more revealing was the adolescent sexual activity and pregnancy rates in the area's most rural communities. In a programmatic response to this information, the following information was included in a submitted grant proposal:

> Two-thirds of all schools in the United States are rural and three-fourths are small. National statistics reveal that most teenage births are from small school systems, and yet, rural communities most frequently overlook the problem of teenage pregnancy. Tragically, some cultural factors and the resource deficiencies of these communities actually facilitate early pregnancy. (Helge, 1989)

Not only has teen pregnancy been described as an epidemic in rural America, but there has been a lack of studies or needs assessments to learn more about this issue. The following additional material from the same grant proposal highlights this point, as well as LFSCC's emphasis on the issue:

> Of all the youth cultures researched and addressed over the years, rural culture has rarely been recognized, much less understood. Significant attitudinal differences have been documented between rural and urban adolescents, leading researchers to suggest the need for separate curriculums in family living to meet the differing needs . . .
>
> Seldom has any active needs assessment been conducted in these under-served communities even though adolescent sexual activity and pregnancy rates reveal desperate needs. As a result, no curriculum to date addresses the culturally sensitive issues known to rural communities. (2–3)

The Lawton/Fort Sill Community Coalition began its Oklahoma Rural Abstinence Project to address this deficiency. The goal of the project is to create a curriculum designed to reduce the proportion of rural adolescents who have engaged in premarital sexual activity, including but not limited to sexual intercourse; to reduce the incidence of out-of-wedlock pregnancies among rural adolescents; and to reduce the incidence of sexually transmitted diseases among rural adolescents. The project conducted needs assessments through focus groups and community discussions in four rural communities. Based on the information gathered, a culturally sensitive, abstinence-only educational curriculum was developed.

Kiowa County was selected to participate in the project because of its largely low-income rural population, racial diversity, and lack of sexual abstinence education. The research team's visit to Kiowa County brought us to Hobart, the county seat. The visit left us with the general impression of a town that is physically isolated, with few structured activities for youth outside of school. There are no major shopping centers and no movie theater in Hobart. For such things, youth have to go to Lawton (64 miles away) or Altus (35 miles away). When the focus groups for the needs assessment were conducted in areas like Kiowa County, girls were highly vocal about their strong desire for close relationships and the emotional component of physical intimacy. In areas where there are few alternatives, it is not surprising that girls and boys would seek connection and meaning through physical and emotional closeness, which is why the prevention of early sexual activity might be particularly challenging in rural communities.

The Contribution of the Lawton/Fort Sill Community Coalition to the HC • HY Movement

The Rural Abstinence Project provides an interesting lens for LFSCC's focus on reaching a broad range of youth and blending prevention and strength-based programming. Rural youth were active participants in providing the information needed for the abstinence curriculum, which clearly has a prevention focus. Of the eight initiatives in our study, LFSCC has been perhaps the most successful in addressing diversity at the participant, leader, and sector levels. The Rural Abstinence Project provides one more example of that emphasis.

The Lawton/Fort Sill Community Coalition has demonstrated that the use of both prevention and strength-based models can be integrated and might even be optimal, at least within certain contexts. In the coalition's work, the two approaches are seen as complementary rather than contradictory. The leadership team of the coalition is professionally experienced in prevention and youth development work and is savvy about speaking the prevention and Developmental Assets languages equally well. Their ability to "code switch"—to speak the language of prevention or Developmental Assets when needed—and to see this back-and-forth dialogue as natural and essential allows them to promote their multifaceted programming agenda quite seamlessly. Their bilingualism in this regard also allows them to take calculated risks in pursuing available resources and community partners. Rather than simply pursuing pots of money, the coalition is exceptionally mindful of the long-term consequences of its fund-raising and program development decisions. As opposed to feeling locked into a direction, however, coalition leaders have selected an approach that might be termed strategically fluid. They can shift directions to address specific community needs without compromising their prevention and strength-based identity.

Finally, it is critical to note that LFSCC is particularly *youth minded*; that is, coalition leaders seek the input of a broad range of youth as they develop their approach to the work. This mind-set may be especially important when drawing on a prevention orientation, which can feel to youth like a "just say no" adult-centric attitude. For prevention messages to be effective, it is important that the community's young people endorse them. Our code *Doing With, Not Providing For* is applicable here and is pivotal to sustained and broad-based youth representation. Adults can either make youth the passive recipients of services or programming—*providing for*—or promote their constructive and active participation in activities and leadership positions—*doing with*. A participant in a staff interview shared that

> we're not the type that sits there and says, "Well you should do this or you should do that." We let them make their decisions, okay? And we try to interact with them and let them solve their own problems, you know? [But we] give them the tools to do it ... Back in the olden days we used to think we had to do the program, and then we'd say, "Here. This is for you." Well, that didn't go over very good. So then we finally got smart and wised up and decided, "Hmm, maybe we ought to let them be in on the decision making and the process. And then they would take ownership. And then they would stay involved."

Recognizing that there were no youth representatives on boards in the city of Lawton, LFSCC and its partners organized a team of youth and adults to choose leadership projects and to gain youth representation for these projects on the city council and school board. At the time of our study, the success of these efforts was unknown, but the efforts themselves illustrate the types of proactive relationships among adults and youth that help define the demographic and political nature of representation within LFSCC.

It is the unapologetic and comprehensive blending of prevention with Developmental Assets that makes the Lawton/Fort Sill Community Coalition such an important contributor to this study. As the strength-based movement has strived to gain momentum nationally, there can be resistance to approaches like this one. It is important to remember that the national HC • HY movement is a risky and courageous response to long-held, deficit-oriented approaches to understanding and responding to young people's needs. For the movement to retain its vitality, it must be open to its own assumptions being called into question. LFSCC provides just that sort of questioning, and does so with its focus clearly on the needs, interests, and talents of local youth.

Postscript

The Oklahoma Institute for Child Advocacy remains a key "go to" place for asset-building ideas and information. According to Director of Youth Initiatives Sharon Rodine, "We incorporate asset-building language, resources, and activities into all of our youth-related programming and initiatives."

Chapter 9
"Leaderful" Communities: The McPherson, Kansas, Tri-County Asset-Building Initiative

If we can identify those things, even if we're doing them well, if ... we know what they are, ah, maybe we won't let them ... fade away. (Initiative staff member)

The asset-building efforts of McPherson, Kansas, and its partnership with the three counties of McPherson, Reno, and Harvey has been shaped by a statewide emphasis on supporting the development of local community leadership. (Consistent with local references to the McPherson Tri-County Asset-Building Initiative, we use the terms "McPherson," "McPherson initiative," and "McPherson tri-county initiative" interchangeably to represent the larger partnership.) Some Healthy Communities • Healthy Youth (HC • HY) initiatives, such as Portland's Take the Time, emphasize a grassroots approach to mobilization on behalf of Developmental Assets. Others, such as Traverse Bay, Michigan's GivEm40, prioritize the engagement of established community leaders. Asset building in McPherson presents an interesting juxtaposition to these orientations through its emphasis on leadership training for both new and experienced leaders. McPherson's asset-building initiative emerged from a belief that a community's residents require specific skills in assuming effective leadership roles, and cannot simply rely upon life and career experience to successfully mobilize and produce social change.

The story of the McPherson tri-county initiative illustrates the impact of skills training on the initiative leaders' strategic actions. As the only initiative whose Developmental Assets work originated and remained rooted in the business sector, McPherson's profile also depicts creative links between the Chamber of Commerce and the promotion of children's healthy development. Further, the initiative's origins in the business sector seem to reflect a different approach to securing financial sustainability than those embraced by initiatives originating in human service or education sectors.

Although this community's leaders say formal asset building is in its "infancy," the intentional, positive support of youth strengths is seen as a longtime tradition in McPherson and the surrounding area. The promotion of Developmental Assets is viewed more as an affirmation of what people already know and do, rather than a shift from thinking about youth in deficit terms. Although not formally affiliated with Search Institute's national HC • HY initiative, McPherson's tri-county

M.J. Nakkula et al., *Building Healthy Communities for Positive Youth Development*, The Search Institute Series on Developmentally Attentive Community and Society 7, DOI 10.1007/978-1-4419-5744-3_9, © Springer Science+Business Media, LLC 2010

asset-building work serves as a model for the Kansas statewide initiative supporting the development of local leadership. The study of McPherson's approach provides an opportunity to consider the role of formal leadership training as a starting point for community-wide asset building. In particular, formal leadership training may provide a specific skill set, such as strategic planning, assessment of readiness, and engagement of key individuals and constituencies without which an initiative will struggle to progress.

McPherson's staff stressed the necessity of our visiting the surrounding towns in the tri-county area, reiterating throughout our visits that the initiative "is not just about McPherson." Like the asset-building efforts in Oklahoma (Chapter 8) and Nevada (Chapter 10), McPherson faces the need to balance the use of prevention and strength-based youth development models. Reno County, one of the partnering counties in the McPherson initiative, has a deeply entrenched history of prevention funding and utilization of the widely acclaimed Communities That Care prevention model. Reno County's asset-building leadership was adopted by the public health sector, a sector with a long tradition of a prevention orientation that relies on state and federal grants for financial sustainability. The tri-county initiative leadership, on the other hand, with its roots in the McPherson Chamber of Commerce and the community's thriving business sector, typically raises funds through private donors, rather than competing for grants.

The McPherson Chamber of Commerce prides itself on fostering a quality of life more typical of small towns and cities of an earlier era, with a particular emphasis on the importance of family, faith, and community participation. Support for healthy youth development is viewed as crucial to sustaining this quality of life, especially within the area's modern landscape of giant chain stores, fast-food restaurants, and dwindling agriculture. McPherson is an award-winning small town recognized by Main Street USA, part of the National Trust for Historic Preservation, which advocates for the revitalization and historical preservation of commercial downtown areas in cities and towns across America. The Chamber of Commerce promotes McPherson as "The Hometown You've Always Wanted."

The focus on revitalization has posed a conflict of competing needs in McPherson. Due to the poor condition of the storefront that housed it, the downtown Teen Center was closed during the months between our first and second visits to the area. We had been interested in visiting the center since it was used primarily by youth from lower-income families. The closing of the Teen Center is emblematic of a challenge confronted by all the communities in our study: that of securing the involvement of those distressed or marginalized youth who are less easily engaged in asset building.

Community Features: Natural Resources and Foresight

McPherson's central location and easy highway access to Wichita, Kansas City, and Oklahoma City make it a small town with an urban sensibility. The town of

McPherson has a population of 13,770, nearly half the population of the larger county, which has 29,554 residents. McPherson County includes eight distinct communities, and the asset-building effort includes these communities plus many of those in the adjacent counties of Reno and Harvey.

Several contextual features are relevant to an understanding of the emergence of formal asset building in McPherson. One is the presence of natural resources, such as the Equus Beds Aquifer, which provides a consistent source of water and oil to the area. The aquifer was discovered during the 1930s when the national economy had plummeted into the Great Depression. Another contextual feature is the legacy of proactive and entrepreneurial leadership evidenced in decisions made to diversify McPherson's economic base so that it was not solely dependent on agriculture. The focus on preservation of natural resources and innovative thinking on the part of leadership are reflected in the mission and role of the Chamber of Commerce. McPherson is one of only two community initiatives in our study to have the business sector actively championing positive youth development (the other being St. Louis Park, Minnesota). The involvement of McPherson's thriving business sector has been promoted through the leadership of the Chamber of Commerce. To adequately understand the emergence of the strength-based tri-county initiative, we needed to learn about the Equus Beds Aquifer, "The Electrical Power Story," and the mission and function of the Chamber of Commerce, specifically, the Chamber's model membership in the Kansas Community Leadership Initiative.

Competing Resources: Oil and the Aquifer

When we first visited the community in November 2002, McPherson was in heated debate over whether to permit a foreign corporate dairy to locate in the area. Over the past 60–70 years, the community has been wisely entrepreneurial in attracting outside business to McPherson through the use of its oil reserves, which has offset potential overreliance on agriculture. In this case, however, the recognition of potential opportunity was offset by concern over a large dairy's impact on the area's water supply, which brings us to the Equus Beds Aquifer.

Equus Beds is a large area of loose, fertile soil sitting upon a type of rock that holds water, akin to a natural sponge. The geology of the high plains aquifer (natural water supply) system includes a layer of porous rock that becomes filled with water from precipitation and from streams and rivers. The vertical thickness of the aquifer ranges from 50 to 250 feet. Bedrock seepage, precipitation, and streams and rivers recharge the pore spaces of the rock. The thin soil covering the aquifer makes it vulnerable to surface activities such as waste management from both industry and agriculture. Contamination can come from many sources, such as oil and gas drilling and transport, industrial wastewater, animal feed, storage leakage, pesticide use, solid waste, and hazardous household waste. Farm chemicals, used to enhance soil productivity, kill pests, and control weed growth, are also a source of pollution. We learned that, in general, chemical application, runoff, and wind transport are the main ways in which water in rural areas becomes contaminated.

Due to its dependence on precipitation, the Equus Beds Aquifer is carefully monitored and protected from overuse. Each county has agencies that are part of the state's Department of Health and Environment, which regulates groundwater use and monitors sources of identified and potential contamination. Some of the strategies involved in rural pollution management include reduced pesticide application, timing of applications, and crop rotation. Bringing a large corporate dairy farm into the area risked upsetting the delicate balance of economic prosperity and natural resource preservation reflected in the concern over the Equus Beds. A natural resource of this importance to the region, combined with the additional resources of native oil and rich agricultural land, creates economic and quality-of-life tensions that require tremendous wisdom to manage properly. It is little wonder, then, that McPherson's asset-building initiative was constructed around leadership training and guided by the Chamber of Commerce, the community organization responsible for keeping the area's resources in balance.

The Power Utilities Story: Entrepreneurship and Risk

> Amid the 1960s agricultural boom, Mr. Diehl and other McPherson leaders saw a dark future for farming. Consolidation was prompting some farm families to sell. These families, though flush with cash, saw no chance to stay in McPherson ... Town leaders set out to change that. They developed, over time, three industrial parks and issued $426.8 million in industrial-revenue bonds—an astounding amount for a town this size—giving manufacturers lucrative property-tax abatements. (Kevin Helliker, "One Kansas Town's Winning Power Play")

One of the factors contributing to McPherson's robust economy was an inventive business proposal made during the 1970s that ultimately secured one of the lowest-priced electricity rates in the country for both residents and businesses. The proposal was to construct and maintain a gas-fired power plant solely for the use of Western Resources, Inc., a private utilities company based in Topeka. McPherson's oil reserves allowed it to develop gas-fired electrical capacity, which Western needed to meet peak load periods. It was not economical to build additional coal-fired plants for peak operation only. As a municipal utility, McPherson could use tax-exempt financing to build and operate a gas-fired power plant solely for Western's use on an as-needed basis. In exchange, McPherson would receive low-priced electricity from Western's coal-fired plant for residential use. The McPherson City Council approved the proposal for the plants, which, if it failed, could have left McPherson in debt from borrowing the money to build the new plants, and with triple the electrical load it needed.

The venture was successful and has since attracted energy-intensive businesses to the area, which has served to maintain high employment levels even during national economic downturns. In fact, the presence of energy-intensive factories as part of a diversified industrial base has supported a robust local economy at a time when many small midwestern agricultural towns like McPherson "have suffered slow deaths," as a member of McPherson's Chamber of Commerce put it. Since 1960,

McPherson County has built three industrial parks and attracted 49 manufacturing businesses, including CertainTeed, Johns Manville, Abbott Laboratories, and Ferguson Production. These companies employ more than 4,000 people.

The infusion of new industry was further stimulated by start-up loans that were made available to assist companies in taking advantage of the low energy costs incentive. The appeal of settling in a small town with ample employment opportunities, high-quality public schools and local colleges, and little crime resulted in McPherson's becoming a highly desirable place to live. The McPherson Chamber of Commerce has been instrumental in cultivating the community's success and strategic in enhancing it. The skilled planning and negotiation reflected in the power utility and dairy farm—Equus Beds stories are more than guiding metaphors for Developmental Assets work. They represent a way of being in McPherson, a way of being a community leader.

A Chamber of Commerce for Business and Youth Development

As noted earlier, the McPherson Chamber of Commerce is the lead sector advancing Developmental Assets work in this tri-county area. The Chamber provides expertise in networking, marketing, strategic planning, fund-raising, and communication. Its proficiency in these arenas underscores the capabilities a business sector can contribute to the development of a successful HC • HY initiative. The connection between the economic health of a community and its children's healthy development is not typically an active concern of a Chamber of Commerce, however, even though rhetoric might suggest otherwise. With the McPherson Chamber's active presence, the community's asset-building approach reflects many of the attributes of a thriving business combined with an activist's commitment to community-based promotion of healthy youth development. This combination manifests in the cultivation of a "leaderful" community, one in which leadership is skilled, proactive, and plentiful. The story of asset building in McPherson is very much one of the synergistic interaction of civic leadership development and business acumen.

The McPherson Chamber of Commerce has approximately 520 members and 220 associate members, which includes representation from the retail industry, service sector, nonprofit organizations, professional organizations, and manufacturing. The Chamber of Commerce staff includes a president, executive vice president, administrative assistant, and office assistant. The unofficial leader of the asset-building initiative in McPherson is the Chamber's executive vice president. The Chamber of Commerce shares a downtown office building on the main street with the McPherson Convention and Visitors Bureau. Our base of operations during both of our visits was the Chamber building, which had a conference room we could use for focus groups, as well as individual offices for conducting interviews. The reception area of the Chamber is noteworthy for its child-friendly atmosphere. There are children's drawings, handprints, and height measurements prominently displayed, along with asset-building information, and the typical maps, brochures, and magazines describing area attractions. The executive vice president's office has glass walls, which allows her to view arrivals in the reception area as well as underscores her own visibility and, to a degree, availability.

The Chamber's mission, according to its Web site, is "To work as an organization of industries, businesses, and professionals to enhance the economic, civic, and cultural interest of the McPherson area." Some of the stated goals of the Chamber are to develop a comprehensive street and highway program; to recruit, train, and retain a quality workforce; to work with city and county agencies to ensure affordable housing; and to inspire and develop local servant leadership (i.e., leaders who can effectively serve their communities) to develop an informed electorate. Partnering organizations include McPherson Industrial Development Company, McPherson Main Street, and the McPherson County Small Business Development Association.

Many of the Chamber's activities are intended to stimulate and promote business development by attracting people to McPherson. Those activities include hosting orientations for new businesses, ribbon cuttings for grand openings or remodeling efforts, arranging industry tours, providing statistical and community data, sponsoring seminars, and hosting annual events. Several community events hosted or supported by the Chamber of Commerce were referenced during our visits to McPherson, such as "When Pigs Fly," "Farm Forum," "Rural Appreciation," the Turkey Creek Golf Tournament, and, most relevant to our study, "All Schools Day," a McPherson tradition originating from a 1913 ceremony honoring county graduates. It is now an event that brings 30,000—35,000 people to McPherson for a 90-minute parade. As newcomers discover, it is a McPherson holiday, and businesses are closed to honor and celebrate all schoolchildren past and present:

> When we first moved to town and experienced our All Schools Day parade, I'd never experienced anything like that in my life, where all of the schoolchildren from nursery school all the way up through the high school are on parade ... We stand at the sidelines and we cheer them ... And it's been going on for 80-some years, you know ... it's moving. I ... cried. Yeah, just stand on the sidelines, just experience that ... rich heritage and ... affirmation of children, of education, and the whole county stops. Businesses shut down. And they all come out and celebrate children. (Focus group participant)

All Schools Day dramatically illustrates the community's affirmation of children and education. There are three small colleges in McPherson, which in a town this size also reflects a strong educational ethos. The colleges bring a new population of students and their parents to McPherson each year, which energizes local businesses and introduces the community to potential residents. Participants in our focus groups stated that there is a history of college graduates returning to McPherson to raise their families, including those who attended one of the local colleges after growing up in the community only to move away temporarily after graduating:

> You saw people moving back after they'd grown up—which was unusual, even then [decades ago], in a small town. People generally didn't come home to a town of ten thousand and raise their family after having grown up there. And that wasn't unusual in McPherson then, and it's not unusual now. (Focus group participant)

The connections between education and quality of life in McPherson are not lost on the Chamber of Commerce, which uses print, radio, and local cable TV to reinforce the connections and build community loyalty. The Chamber's executive vice president writes a column on local activities in the *McPherson Sentinel* as well as hosting a weekly radio program. Through both of these outlets, education and developmental assets are commonly featured.

Developmental Assets and the Chamber of Commerce's Mission

> For those who question the Chamber's involvement with youth issues, we remind you of the economic benefits. Healthy children require fewer dollars for deficit issues like truancy, delinquency, addiction, and teen pregnancy. Healthy children learn better, become better workers and better citizens. If we connect with them now, they will connect with us later. They will remain here, work here, and invest here. (Chamber of Commerce column in the *McPherson Sentinel*)

Community events, such as quarterly breakfasts, pancake feeds, and annual dinners are used to promote positive youth development. A Chamber-hosted dinner invited Judge Tom Webb, a Korean American who was adopted by a White American family at age seven following the Korean War, to deliver a speech titled "How to Make a Positive Difference in a Negative World." Webb shared personal experiences of adults who had made a difference in his life as a child. Part of the Chamber's asset-building campaign was to ask grown-ups to reflect on the adults who had made a difference in their lives as children. Stories from these reflections were printed in the *Sentinel.*

We had an unplanned opportunity to observe the monthly "Ag[riculture] Related" subcommittee meeting of the Chamber of Commerce. The meeting of this group of some 30 men involved in farming began with a debriefing of a recent "Farm City Day." The discussion began with members voicing the need to more proactively expose young people to and mentor them in the work of farming. A decision was made to contact the high school dean to encourage coordination with the Future Farmers Association. There was also discussion of some of the obstacles to more actively engaging middle school students interested in agriculture. Unlike high school students, who have more access to transportation and school organizations, such as the Future Farmers Association, the group felt that opportunities needed to be developed to further expose the middle school youth to modern-day farming. Discussion about ways to enhance youth involvement in agriculture was a surprising start to this meeting, given the controversial issue facing the community, namely, whether the natural resources of the area could sustain the foreign corporate dairy hoping to relocate there.

The Chamber not only "walks the talk" of being youth minded itself, it also invites businesses and other organizations into a larger community partnership through which they can explore strategies for contributing to positive youth development that they might not consider on their own. The investment of the entire community is needed, according to the Chamber leadership, if the health and positive development of all children in McPherson is to be promoted. A leadership focus group participant stated:

> While we have a good community, we can be better. And part of making [us] better is ... the generation that comes behind us ... This is the best place to invest our time and our energies, and if we can make them healthy, then we really will secure the quality of life for future generations in McPherson. And we're starting to see [a] change [in the level of support for youth]. But it's got to be bigger yet—because it's been invisible for so long. We've got to bring it out from behind the walls of the Teen Center, and we've got to bring it out from behind the juvenile corrections program and the courts. And it would be, I think, to our advantage, and to our state's advantage, ultimately, if we can make asset development the basis for juvenile corrections.

The Chamber's work is thoroughly infused with the larger mission of supporting local civic engagement, educational opportunity, and community service. The "strategic initiatives" of the McPherson Chamber of Commerce, according to its Web site, include "develop[ment of] an informed electorate" on "major issues of public policy." This task is consistent with the Chamber's active involvement in the statewide leadership initiative noted earlier. The Kansas Health Foundation (KHF) sponsors the Kansas Community Leadership Initiative in an effort to promote leaders who can effectively serve their communities. McPherson's Chamber of Commerce has been an exemplary member of this initiative and as a result has received a substantial amount of financial support from the KHF for its asset-building activities.

The Kansas Health Foundation: Servant Leadership and Children's Health

> For a philanthropy, which exists to improve humankind and to make change possible, the bottom line is that leadership development is human development. It adds value to the financial resources provided by the Foundation, offering human resources to "get the job done." (KHF Web site)

The guiding mission of KHF is to improve the health of all Kansans. This mission has evolved toward a view of health that includes social determinants such as jobs, education, and the environment. KHF takes an interesting position on healthy development, framing it as an artifact of feeling in control of one's life. Working from that sense of control, the foundation argues, citizens are better positioned to take care of themselves and effectively contribute to others and to their communities:

> Whether a person is healthy or not is related to how much control he or she has in their lives, particularly in work environments. Leadership development can offer individuals a view of their world that allows them to maintain a sense of control they otherwise might not have. (KHF Web site)

In conducting listening tours of the state in 1988, and again in 1995, KHF found that leadership "surfaced as both a barrier and a potential strength when addressing the complex health issues facing Kansans" (KHF Web site). In 1988, the issue was lack of leadership, and by 1995 the issue was "too few leaders asked to do too much." Leaders were described as lacking financial and human resources, and the specific skills needed to promote community sustainability. In 1999, KHF created the Kansas Community Leadership Initiative and invested funding in creating a leadership curriculum. The curriculum uses interactive classes to address collaboration, consensus, visioning, and empowerment. A central component of the approach is the training of others throughout the community in an effort to build a highly skilled leadership base. Focus group participants shared the following perspectives:

> The Kansas Health Foundation's initial premise ... as far as impediments to making a healthy place for children ... is that ... leadership [is the key]. Where it [community is thriving it is the result of leadership. But where it's not thriving, it's a lack of leadership.

> What they have done extremely well is to allow us to continue to expose other people to what they're about. And I know through the leadership institute for the last several years, we've had an opportunity to take other people as our guests, so we do talk about asset building and [we have] an opportunity to hear a variety of speakers. And I think we've grown our group through that ... availability, that participation. Plus, [we've] enhanced the awareness and commitment of those who are involved through that program and those opportunities. I would hope that they would continue to do those kinds of things, and allow us to bring other people with us, particularly young people.

The Kansas Community Leadership Initiative is a program that provides intensive training in "community visioning, group facilitation, servant leadership, learning styles, collaboration, consensus building, and community history timeline development" (KHF Web site). The Leadership initiative sponsors an annual three-day conference that both recruits new leaders and offers ongoing training for alumni of the program. This annual conference has grown a statewide leadership network. In preparing citizens to become local leaders, the Leadership initiative is designed to build the social networks that sociologist Robert Putnam claims have deteriorated across America in his immensely popular book, *Bowling Alone: The Collapse and Revival of American Community.* Participation in the Leadership initiative strengthened links between McPherson's leaders and those in nearby Harvey and Reno counties who were either using Developmental Assets, or had already become familiar with them. These three counties have now collaborated on several projects and make up the tri-county asset-building effort centered in McPherson.

Having adopted a public-health orientation suffused with attention to the social determinants of health, it is not surprising that KHF decided to roll back its focus on adult health to earlier phases in the life cycle:

> If children could grow into caring, contributing, thoughtful and tolerant adults, they would be healthier—both in terms of physical and emotional well being, as well as make better decisions and care for their communities well into the future. (KHF Web site)

In the mid-1990s, KHF convened national and local experts to consider the role of philanthropy in promoting children's health. Peter Benson, president of Search Institute and a native of Kansas, was one of the early consultants. Search Institute's research on the factors that positively influence children's health became the scientific foundation of a $2 million social media campaign, "Take a Second, Make a Difference," begun in 2001:

> It was believed that by creating an environment, the focus would be on challenging and improving all those influenced that surround a child. Leadership development—particularly in communities—and communication plans such as media campaigns were approved as appropriate strategies to begin to create this environment.
>
> As a result of numerous discussions, Board members examined the role of "pivotal moments" in their own lives and realized the bottom line goal for the Foundation's work should be a caring adult for every child. (KHF Web site)

The linking of leadership to children's health found an ideal home in McPherson. The Chamber of Commerce proved particularly adept at applying the social marketing of the Take a Second, Make a Difference (TASMAD) campaign through its newspaper column and weekly radio program. In 2001, McPherson High School was chosen to pilot a second phase of the campaign—"TASMAD Goes to School"—which specifically focused on adult–youth and peer relationships within school communities.

Two leaders, in particular, have been instrumental in linking business and health in the service of asset-building efforts in McPherson. One is Dr. Greg Thomas, a family practice physician, who spearheaded efforts to establish McPherson as one of six regional Medical Education Sites for University of Kansas medical students. The other is Cheryl Lyn Higgins, the executive vice president of the Chamber of Commerce. Together, Higgins and Thomas crafted a strategy that would link children's health, business, and local leadership as a critical first step to mobilizing the community on behalf of Developmental Assets. Locating a Medical Education Site in McPherson established a structural connection between the health and commerce sectors, given that funding for selected Medical Education Site activities would be managed by the Chamber of Commerce in its role as the sponsor for local asset-building initiatives. Higgins recalled:

> Because the Health Foundation is at the center of much of this ... the first thing that happened was that they offered funds to the University of Kansas for primary care sites throughout the state. And for the South Central site, McPherson was invited to consider hosting that, as were a number of communities. And we were surprised that not as many communities responded to the invitation. We did. We were very aggressive in pursuing the site.

However, Higgins described how she still needed to explicitly link the work of the Chamber of Commerce to children's health and asset building:

> While we loved the concept of asset building ... it was hard for me to figure out how we could embrace that as a Chamber initiative. But I was still on the advisory board of the Medical Education Site and one day Dr. Thomas shared with us the fact that the Health Foundation, the primary care physician program, was being offered an opportunity for a $20,000 grant for a community project. Suddenly the lights went [on], you know? And I said, "Have I got an idea for you. Why don't we use that money to promote asset building in the community?"

When KHF provided the $20,000 grant to Memorial Hospital, Higgins followed up with her recommendation to design a program to introduce assets-based youth development to the region as part of the Chamber of Commerce's mission:

> When the hospital Medical Education Site accepted [the asset-building initiative] as a project, that gave me a way to bring it into the Chamber—because then I could go to the board and say, "Our partner, Memorial Hospital, wants to embrace asset building as a project. And we have this opportunity to work with them." I remember the exact words of one of our board members at the time: "How can we say no to the hospital? And how can we say no to children?" So that allowed us to incorporate [asset building] as a Chamber initiative, and allowed me to officially give my time to developing [an asset-building] conference.

Key Activities and Achievements

Connecting the Counties

With Dr. Thomas's blessing, the Asset Advisory Board was set up within McPherson's Chamber of Commerce to plan for the use of the Medical Education Site grant. Shortly afterward, the board was merged with the Chamber of Commerce's existing Education Committee to form an Asset Education Committee. The Medical Education Advisory Board believed that the Asset Development Project should be a tri-county initiative. Higgins, along with other members of the board, recruited key leaders from Reno and Harvey counties to join with a core group from McPherson to launch a tri-county project. Dr. April Osborn, who is the director of a community health coalition in Reno County, and Kristi Bohling-DaMetz, who is linked with the Harvey County Regional Prevention Center, joined forces with the McPherson group to organize the Tri-county Children's Conference in 1999. Dr. Osborn was already familiar with Search Institute's Developmental Assets framework and had efforts under way to integrate health and development programs in Reno County. Ms. Bohling-DaMetz had been introduced to asset building through a family member involved in a Michigan initiative supporting strength-based youth work. Higgins, Osborn, and Bohling-DaMetz all drew from their networks and convened a program of dynamic speakers to launch the Assets initiative.

The theme of the conference was "A Little Effort Makes a Big Difference," an adaptation of the Kansas Health Foundation's "Take a Second, Make a Difference." Higgins recruited Bohling-DaMetz and Gary Monford, of the Dillons supermarket chain, to join KHF's Community Leadership Institute. Support from Dillons was instrumental in financing the next strength-based tri-county project: hosting Erin Gruwell and the Freedom Writers. Higgins and others had been introduced to Gruwell and her work through the Kansas Community Leadership Initiative. Erin Gruwell is an award-winning high school teacher from Long Beach, California, who inspired her "unteachable, at risk," and largely minority students to explore their experiences of oppression and discrimination through the study of autobiographical narratives, starting with the diary of Anne Frank. Gruwell's students went on to write their own stories and publish them as *The Freedom Writers Diary: How a Teacher and 150 Teens Used Writing to Change Themselves and the World around*

Them. Profits from this book now help pay for these students' college education. The Tri-County Asset initiative was able to partner with an endowed lecture series program in the region to bring Erin Gruwell and two of her students to speak to students in all three counties. A focus group member commented:

> While we're the McPherson Chamber of Commerce, our hope was that this really wasn't going to stay in just McPherson. When we counties reconnected to bring ... the Freedom Writers to the school systems here ... we didn't just look at that as an opportunity for the city; we looked at it as an opportunity for the counties and found ways to get the ... schools in the area [involved].

Tri-County Differences: The Emergence of Paradigm Clashes

> The things that we're doing, the ways we're sustaining our efforts ... are different ... but we still come together. We still share ideas, and, if possible, resources and presenters. But we still have maintained our own unique characteristics. But the one thing, the overriding thing, that has kept us [united] ... is the sharing fashion that made this happen. We may not agree on how it's best reached or all of those aspects, but we ... firmly believe that this is the way to make our communities healthy—not just for children, but for adults, too. Because, obviously, if you make it a great place for children, you're making it a great place for adults as well. (Tri-county focus group participant)

There are distinct differences in how asset-based efforts are designed and implemented in McPherson, Reno, and Harvey counties. At the center of these differences is the issue of whether programs are defined solely as asset building, or whether a blended model with both strength-based and prevention components is followed. While McPherson County's asset initiative is based in the Chamber of Commerce, Reno and Harvey counties' asset programs are located within health and human service sectors. Whereas McPherson has cultivated the financial support of the business community, Reno and Harvey counties rely largely on prevention monies, which can be perceived as deficit based. Reno County's Dr. Osborn summarized the paradigmatic conflicts: "If a strength-based model works, it rejuvenates itself. A deficit-based model continues to have need, you continue to have to fix the problems."

In our focus groups, the county leaders debated their concerns concerning future funding of strength-based work. Dr. Osborn predicted a challenge to McPherson's efforts as they sought to engage traditionally deficit-based community sectors, such as juvenile justice and child welfare, in their asset-building initiative:

> I feel sure that, when you get to a certain level of ... prominence in the community, you're going to have the same clash that we have. Because you're gonna have your juvenile justice people funded in that way, and it's gonna be the same thing [we experience].

The McPherson leadership countered the concern with examples of having successfully engaged these sectors:

> We haven't had that experience of clashing ... In fact, we had representation from juvenile justice in the tri-county conference as participants, and we had several who continued with

> us in the Asset Speaker's bureau. So, we haven't seen that real clash over resources or [philosophy] . . . at this point . . . And that may be because we're not asking for any funds.

Reno and Harvey counties are not alone in tackling the "balancing act" of financial dependence on "deficit funding" for strength-based work. Several initiatives in our community case study are wrestling with how to negotiate these paradigms in a way that uses the strengths of each orientation. The McPherson initiative has not pursued prevention monies through state or federal grants, resisting what it perceives would be "selling out" to a deficit-oriented philosophy in order to further its work. In fact, there is a strong concern that accepting prevention money would inhibit the initiative from pursuing its leadership-based approach to asset building. This "prophecy of the golden handcuffs"—being tied to specific funding requirements was viewed as potentially restricting enthusiasm and leadership creativity. As with their concerns over the Equus Beds Aquifer in the face of corporate dairy money, the McPherson Chamber of Commerce prioritized a longer-term vision of community health over immediate payoffs.

An obvious exception to using government or foundation funding for its asset-building work is McPherson's relationship with the Kansas Health Foundation. As described earlier, McPherson has been recognized as a model leadership community by this powerful funder, and Chamber of Commerce's vice president Higgins has been described as "the darling of the Foundation." In an interview she stated the following:

> And some of the . . . pressure from the Chamber perspective is to keep a balance . . . because we have so many other initiatives . . . you can't allow one to become more dominant . . . you know, dominate your time over the others . . . And the other thing . . . I have really tried to work to do is to reduce the perception that we're owned by the Kansas Health Foundation.

Although affiliation with KHF in no way counters McPherson's asset-building philosophy, the foundation's powerful presence still raises concerns over perceptions of autonomy. The more powerful the partnerships, it might be argued, the more powerful the organizational interdependence. In general, this would seem to be a positive characteristic for collaborative work; at the same time, it is critical that interdependence not interfere with innovation. This caution applies to the county relationships as well. Following the initial tri-county focus, the McPherson initiative subsequently promoted additional projects tailored to its more immediate community. One event was an asset-based Family Fair; another was a community-wide visioning project. It is arguable that the move inward, for at least part of the work, was influenced by tri-county philosophical differences.

McPherson-Specific Projects: Focusing the Lens

McPherson's Family Fair was held in March 2003. Invitations were sent to community organizations, particularly those that were family or children oriented, asking them to participate in a Family Fair focused on the 40 Developmental Assets.

Organizations were encouraged to design a booth with an interactive activity highlighting a specific asset. The response was overwhelming, with more than 40 organizations signing on. The Chamber of Commerce donated its booth to accommodate a community organization, and the Chamber's leadership class of 2002 highlighted the asset of service to others by preparing a meal for all participants of the fair, serving over 800 people. The Family Fair stimulated momentum for the community-visioning effort, Vision 2010:

> One of the weaknesses, I think, in our community and our county is that we don't share the same vision. We've got different organizations and agencies working [in different directions]. And we all know from what we have read and seen that resources are going to be more difficult to come by, that we've got to be more creative, more efficient with the dollars that we have, and more collaborative . . . And I think organizations could do a better job if we all were on the same page, not serving the same role, but knowing exactly what's being done and who's doing it and where we can help one another, or who needs to step aside and not duplicate services and resources and allow somebody else to do it. (Focus group participant)

A "Vision Day" was held just prior to our arrival. Consistent with the Chamber of Commerce's commitment to McPherson's becoming a "leaderful" community, it hired two external professional facilitators to guide them in initiating a county-wide visioning process for the future of McPherson. The first stage of this process, which included conducting some 100 focus groups with residents, had just been completed when we made our first visit to the area in November 2002. In addition to the focus groups, there was also a questionnaire disseminated throughout the community. Vision 2010 is a grassroots initiative focused on identifying what residents—youth and adults—view as the most significant issues facing the community, and designing strategies for addressing them. Five areas were identified as most significant: economic health, child and youth development, shopping, services, and recreation. During both of our visits, we had an opportunity to observe a meeting of the McPherson County Vision Committee, which was composed of "action teams" to address each of the five identified issues.

In our study of the eight asset-building initiatives, we found that efforts made to foster partnership among divergent community sectors with no history of working together is a critical achievement that often gets buried beneath more visible accomplishments. McPherson has been masterful at addressing each of the five areas identified by its residents as most important to the community by fostering sector-, county-, and statewide collaboration. Key leaders support linkages among the Kansas Health Foundation, the Kansas Community Leadership Institute, the local Medical Education Site Advisory Board, and the Chamber of Commerce. The initiative, though still relatively young in its intentional asset-building efforts, has established connections with the media, faith communities, members of the business sector and medical community, senior citizens, the Parks and Recreation Department, civic organizations such as the Kiwanis Clubs, and the public schools and institutions of higher education. Organizations as divergent as the Cooperative Refineries Association and Girl Scouts have mobilized to support the healthy development of youth in McPherson. A representative from one of the oil refineries helps put this collaborative spirit in perspective in a focus group:

> We want to be able to influence the community so that we have a continued pool of prospective employees that meet our needs. But those [influences] turn out to be exactly the same things that are on the assets [lists] ... We want ... citizenship. We want people who think for themselves. So there is just a natural marriage between the things we were desirous of and what this committee is doing ... If you're going to succeed as a business, you have to have a community where you can thrive ... The industry I'm in is oil [refining], and many communities view oil refineries as dirty, nasty and evil places ... How do we become an asset to the community? How do we get youth more involved in looking at [our] business? ... [Youth] get most of their perspectives from the media. Everybody thinks that all company directors are like, you know, the people at Enron.

At the center of this extraordinary sector, diversity and collaboration within the initiative, however, is the core emphasis on leadership development. As sectors are recruited into the work, sector participants receive ongoing training for key leadership skills needed to address their targeted area of concern—skills such as planning, communication, collaboration, and problem solving. It is precisely such skills that allow diverse sectors to work together effectively, according to the McPherson team. It takes more than desire and goodwill; it also takes strategic know-how.

Growth and Preservation: Not "Losing What You Already Have"

> I don't know if it's necessarily a new thing. I think we're ... maybe coming back to some of what we used to be.

> There's some cultural things happening that make it easy to ... to lose what you have. You know, what you had 20 years ago, 10 years ago. (Community focus group members)

We began this profile with McPherson's version of the inevitable tensions between progress and preservation. A community rich in natural resources carries the particular challenge of nurturing those treasures and preserving them for the next generations while it must simultaneously use them for the community's growth and development. In McPherson's case, the coveted Equus Beds Aquifer demanded exceptional care in its service as nature's filter for the provision of the area's fresh water supply. This care was challenged by the opportunities for growth presented by the wealth of a second natural resource, native crude oil, the availability of which provided affordable energy for new industry and residential development. In the cases of the electrical utility plant and the foreign dairy corporation that wished to locate in McPherson, the town's leadership was forced to make decisions that would reverberate for generations. The long history of such decision-making demands creates a fascinating context for the development of the community's Developmental Assets initiative.

Two thematic clusters from our larger analysis most clearly frame our understanding of McPherson's contribution to our study of HC • HY initiatives: *Leadership Wisdom* and *Synergistic Commitment.* Clearly, the McPherson community is rooted in a long history of leadership wisdom. Managing the tensions stemming from its wealth of natural resources has required such wisdom, and the community has clearly stepped up to the challenge time and again. Not only has McPherson accepted the leadership it has had to exercise for environmental and economic reasons, it also has embraced leadership as a guiding tenet for the community's development on a host of levels. Through the support of the Kansas Health Foundation and the Kansas Community Leadership Institute, the McPherson Chamber of Commerce has pursued leadership training for its own staff and for others partnering in the Developmental Assets work. Embracing the Kansas Health Foundation thesis that human development requires leadership development, and that in many respects human development *is* leadership development—learning to *lead* a healthy and productive life—the McPherson Chamber of Commerce promotes leadership training in all of its asset-building activities.

The thematic cluster Synergistic Commitment can be seen as an outcome of McPherson's wise and pervasive leadership orientation. In fact, the approach to leadership taken by the community initiative is intentionally designed to build synergistic connections and commitments. Synergy implies that through interactive and integrative connections, the whole of an entity's efforts is substantially greater than the sum of its parts. McPherson uses leadership training to ignite that synergistic spark. The Chamber of Commerce leadership team argues that individual leaders leading independently are left to carry too much weight and will ultimately underperform relative to a collaborative leadership approach. But collaborative leadership requires broad and ongoing training. It is not enough to want to lead. Leading with one's will and heart will go only so far. Successful leadership requires specific skills training, according to the McPherson and Kansas Health Foundation philosophy. As that training reaches critical community mass, synergistic leadership becomes possible.

The blending of leadership wisdom and synergistic commitment can be analyzed in the McPherson initiative at different levels that may shed light on asset-building processes more generally. On one level, the McPherson partnerships are unusually simple in a particularly beneficial manner. The asset-based work is rooted in one extremely useful context: the community's Chamber of Commerce. This would appear to be an ideal context for catalyzing the work of the initiative in that by design the Chamber reaches out to multiple sectors, including the business sector. One person in the Chamber, the executive vice president, proves to be a determined leader who will champion the asset mission. This leader links the Chamber to the Kansas Health Foundation, solidifying a profoundly influential partnership, a partnership that deepens the Chamber's preexisting orientation toward community leadership development.

The foundational relationship between the Chamber of Commerce and the Kansas Health Foundation brings together three core concepts that will guide the work going forward: health, leadership, and positive child and youth development. Firmly grounded in those concepts, the Chamber is fortunate to cultivate one additional key collaboration: a relationship with a local hospital that becomes a Medical Education Site. As a training site, the hospital accesses a community development grant, which it uses in partnership with the Chamber to begin health-based

Developmental Assets work. Each link in this chain of events proves extremely strong and provides the primary interconnections needed to ground the work and expand it exponentially.

Upon examining McPherson's extensive slate of partnerships and activities from a distance, it might appear that the initiative moved in all directions simultaneously. But this is not the case. The work began in a localized way, focusing on core organizing concepts and collaborations, and grew strategically from there. Every Developmental Assets initiative grows in accordance with its own identity. McPherson's identity is rooted in leadership. It is a community borne of wise leadership and a community initiative bent on breeding such leadership. If leadership development is a core facet of human development, the youth of McPherson County have productive lives to lead. The local Chamber of Commerce believes that to be the case. Nothing in our study leads us to believe otherwise.

Postscript

In the McPherson Chamber of Commerce, the BASIC (Building Assets Successful in Our Community) committee serves as the vehicle to promote the 40 Developmental Assets through events such as Family Fair (every other year), McPherson Wellness Initiative, and OMEGA (Outstanding McPherson Eighth Grade Award). In addition, the committee addresses concerns of educating youth about future options with events such as College and Career Night.

Chapter 10
The Next New Frontier: Nevada's Healthy Communities Coalition of Lyon and Storey Counties

Northern Nevada's Healthy Communities Coalition (HCC) provides a compelling example of integrating asset building with prevention approaches through a complex partnership among six very different rural "frontier towns," as they are referred to by the people in the area. The complexity of this coalition is a testament to the leaders' fierce respect for the identities forged by each of these towns, and to their ability to gain the trust of small communities fearful of being overshadowed or co-opted by larger towns. Each of the six towns has survived challenges associated with isolation and small populations, such as limited resources for its youth, particularly when compared to more urban areas of Nevada, such as Las Vegas. The coalition honors the unique histories of these towns, emphasizing how each of them uniquely enriches the coalition's work.

The Healthy Communities Coalition operates through an infrastructure of six community task forces, each of which proposes asset-building activities to the coalition's larger governing body, which grants funding and shares information across the network. Membership in the coalition requires the leadership of each task force to implement activities within its geographic area, and to coordinate efforts with the other five HCC communities and a larger statewide initiative, which will be described later in this chapter. The challenge of proactively functioning to meet local needs, as well as collaborating at the regional and state levels, underscores the importance of "non-cookie cutter" approaches to asset building: What's best for one community might not be best for a neighboring community, much less a more distant one both geographically and culturally.

Community Context: Grassroots Orientation with Global Vision

The Healthy Communities Coalition is located in Lyon and Storey counties in northern Nevada. As with the McPherson initiative discussed in Chapter 8, the history and geographic location of the initiative profoundly influence the nature of its work. In many respects, HCC is a stunning illustration of the collaborative ethos and grassroots organization characteristic of the rural initiatives in the communities we studied. For sparsely populated communities, separated by vast distances,

M.J. Nakkula et al., *Building Healthy Communities for Positive Youth Development*, The Search Institute Series on Developmentally Attentive Community and Society 7, DOI 10.1007/978-1-4419-5744-3_10, collaboration is a necessity for accessing a range of services and youth development opportunities. However, achieving a successful partnership requires adroit leadership that can establish and maintain a deep understanding and respect for the individuality of the partnering communities. The leadership of HCC has engaged the six different towns and their surrounding communities across northern Nevada into a partnership of some 250 members actively mobilized on behalf of strength-based work supporting youth and the health of the community more generally. An appreciation of the skill and effort involved in crafting an infrastructure that can mobilize a partnership of such diverse communities requires a deeper understanding of the contexts from which the coalition emerged. The structure and operations of HCC vividly reflect features of the historical, geographic, economic, and social development of Lyon and Storey counties.

A Brief History of Northern Nevada Settlement

Two of the communities that are part of the Healthy Communities Coalition, Carson City and Virginia City, were the first areas of the state to be settled. The settlement of the Nevada territory was spawned by the California gold rush during the late 1840s. Eagle Station, now the site of the state capital, Carson City, became a trading post and rest spot for prospectors making their way west, after days of traveling across harsh, desert terrain. The discovery of gold and silver in the Comstock Lode in 1859, in the hills east of the city, ignited further development of Virginia City as a thriving commercial center, and the first industrial city in the West. At that time, the Comstock Lode was the largest silver find in world history. The wealth produced a city with a six-story hotel, mansions appointed with European furnishings, and the first elevator in the West. A host of performing arts were established there, bringing frequent visits from celebrities to the city. The investments made in the Comstock Lode are said to have financed the building of San Francisco. The discovery of the Comstock Lode was also instrumental to Nevada's being granted statehood despite its sparse population. Although there is a common myth that President Lincoln used Nevada silver to ensure the victory of the Union during the Civil War, there is a clearer historical connection between Lincoln's desire to pass abolitionist legislation and Nevada's entry into statehood garnering critically needed votes.

Virginia City, Carson City, and Reno were linked economically and geographically. The settlement and mining required timber for construction of underground mine shafts and for housing the growing population. Water flowing from the mountains provided transportation for the logging industry producing the lumber. The Virginia and Truckee Railroad was constructed in 1869 between Carson City and Virginia City, and later extended north to connect with the transcontinental railroad in Reno. In the present, we will see that Virginia City played a central role in the development of HCC by joining with Dayton, one of the partnering towns, to apply for the grant that began the community coalition.

The settlement of northern Nevada was fueled by economic zeal, and the atmosphere generally has been depicted as freewheeling and somewhat lawless. Crossing the rough terrain required strong entrepreneurial motivation and stamina. The difficulty of traversing the terrain was not conducive to surveillance of activities by the federal government and business stakeholders back East. The isolation no doubt contributed to the fierce independent spirit of the citizenry, and, as such, it is not surprising that an area of the country that historically was not readily observable is the sole state to have both legalized gambling and prostitution.

Juxtaposition of Natural and Commercial Resources

More driving time was spent visiting the six communities of HCC than on any other field visit made in the study. Covering this distance provided firsthand experience of one of the greatest challenges described by initiative members: the many hours spent in the car versus doing the work. Pictures taken of our visit illustrate a striking contrast of images, such as the resort area of Tahoe with its sparkling, glacier lake surrounded by state forest. Another is of Virginia City with its historic downtown, Victorian mansions in the hills nearby, and restored nineteenth-century opera house. A pervasive image is the long stretches of open road cutting through semiarid and somewhat desolate landscape. This is juxtaposed with the passing images of the one-story library in Silver Springs, the state prison, the new high school, and signage advertising the local (legal) brothels. Carson City is a study in contrasts, exemplary of the state. The beauty of the mountain backdrop starkly contrasts with the ubiquitous casinos, a reminder of the danger and allure of a commercialism that had its beginning in the mining days.

A present-day look at Lyon and Storey counties indicates population growth in some communities of the coalition such as Dayton and sparse population density in others. Silver City, the newest member of the coalition, had 12 school-age children in its town population. Each community has different economic bases, such as Amway in Dayton and Amazon.com in Fernley. Silver Springs faces unique challenges associated with the presence of a federal prison and the number of felons who settle in the area following their release. Many people commute an hour or more to Reno or Carson for employment, which presents serious challenges to transporting youth to after-school activities. An additional commuting challenge is raised by HCC's partnership with Raising Nevada, a statewide Healthy Communities • Healthy Youth (HC • HY) initiative, based in Las Vegas. The 8-hour drive south to Las Vegas takes one to a densely populated urban landscape that is described as "another country" by the rural coalition members. In upcoming sections, we will describe the challenges inherent in this partnership of "north" and "south."

The structure and operations of HCC reflect the social, geographic, economic, and political context of northern Nevada. For example, this coalition is unique in our study of Developmental Assets work for its focus on support and preservation of the natural environment. This emphasis is not unrelated to the state economic dependence on tourism. The natural environment of northern Nevada provides a wide array of recreational opportunities, such as big-game hunting, cross-country skiing, fishing, and camping. The beauty of the lake resorts, such as Tahoe, and scenic drives through land inhabited by mountain lions, bighorn sheep, elk, and bobcats serve as a strong magnet for tourism. The physical isolation lends itself to both a rugged individualism in the area's residents, known as "the rurals," as well as a collaborative ethos, since no one small town can financially support its own public resources, such as schools, libraries, human services, health care, and transportation facilities. This collaborative spirit has resulted in the development of strong, long-term relationships among service providers and coalition builders.

Legalized gambling and prostitution are major contributors to the state and local economies. Establishing a social orientation toward positive and healthy youth development in a community economically dependent upon sanctioned risky behaviors is an interesting and unique challenge for HCC, and provides a valuable counterpoint in a study of strength-based community change models. We heard no discussion of attempting to change the realities of legalized gambling and prostitution, but rather of strategies for promoting the health and well-being of youth and adults in communities that must deal with such challenges. When one looks carefully at the combination of resources and risks facing northern Nevada, and the whole state for that matter, it is not surprising that cutting-edge strategies for blending Developmental Assets and prevention approaches are emerging from this area.

Initiative Features: A Nested Network of Collaboration

Three features combine to make the Healthy Communities Coalition particularly complex. The first is the challenge of linking the six disparate communities into a coherent initiative that preserves the uniqueness of each community. Second, the local or northern six-community network is *nested* within Nevada's statewide HC • HY initiative, which raises challenges regarding funding, transportation, and philosophy. And, third, there is the integration of prevention and asset-based work. Any of these three issues would be sufficiently challenging to manage; the combination of them makes HCC a model that requires particular attention for what it can teach the larger HC • HY movement.

The six communities that comprise HCC are located largely, but not exclusively, within Lyon and Storey counties in northern Nevada. In its print material, the coalition describes itself as representing communities in "rural Nevada east of Minden, Carson and Reno" that are "remote ... distanced from centralized services." The six communities are Fernley, Virginia City, Dayton, Silver Springs, Yerington, and Silver City. Each community's task force must be composed of equal numbers of youth and adults. The purpose of the task forces is "to give voice to the community and link the coalition to a grassroots community group," one of the coalition directors stated in an interview. She added that the task force members are viewed as the "local experts" on individual community needs and resources, and that their primary function is to "help prioritize risk factors, promote Developmental Assets and service-learning, implement one service-learning or asset project a year, and participate in the community project proposal process."

Community proposals are reviewed monthly in a coalition-wide meeting, which rotates location in order to share the burden of travel time.

Leadership Roles

Because the functioning of a complex coalition like this one requires particular leadership expertise, any understanding of HCC requires some sense of its leadership structure and functioning. At the time of our visit, the coalition staff included the codirectors of HCC, Christy McGill and Alecia Hoffman, a prevention coordinator, a youth coordinator, and the six task force leaders. The codirectors described the coalition staff as "a group of health and human services professionals and agencies who network, share resources, act as colleagues for those who provide services in Lyon and Storey counties, and . . . provide expertise, support, and feedback to local task forces and the core team."

The primary function of the coalition staff is to provide technical assistance and training to the six task forces in the areas of Developmental Assets, prevention, service-learning, and community building. Facilitating communication among the partnering communities and helping to coordinate their efforts constitute a key focus of the daily staff operations. Care is taken to problem-solve regarding potential resource sharing so as not to duplicate services. Largely due to its reliance on prevention funding, grants acquisition and management is the lifeblood of the coalition, and much effort has been devoted to establishing a culture and infrastructure of accountability, which includes disseminating evaluation materials throughout the northern area of the statewide initiative served by HCC. The mission of the coalition staff includes establishing HCC as a cohesive and visible advocate on behalf of healthy youth and community development at the county and state levels.

The leadership of HCC is predominantly women. Collectively they represent a leadership team that is a blend of youthfulness, dynamic activism, and seasoned wisdom. Christy McGill and Alecia Hoffman share a full-time director position. Their roles entail the larger visioning, progress, and sustainability of HCC. Staff and participants described McGill as a "master coalition builder." Operationally, she has a critically important communication and facilitation role, which includes attending board and task force meetings, and providing direction for the HCC staff. In addition, she manages the Bureau of Alcohol and Drug Abuse (BADA) coalition grant, which required the development of a systematic reporting process for the member communities.

The other codirector, Alecia Hoffman, began her nonprofit work as an assistant to Deborah Loesch-Griffin, a founder of Community Chest, Inc., in Virginia City. Community Chest, Inc., has been a powerful catalyst of youth and family advocacy and resource development in Storey County for 10 years. As a codirector of HCC, one of Hoffman's major tasks is to organize a bicounty youth leadership summit focused on youth development leadership, Developmental Assets, and service-learning. She is also responsible for the implementation of the scope-of-work plans for several grants, and for implementing task force strategies in Virginia City and Dayton.

Pam Ambercrombie implements the Office of Juvenile Justice and Delinquency Prevention (OJJDP) grant activities, with a focus on implementing training and task force strategies in the member communities of Silver Springs, Fernley, and Yerington. She conducts three trainings per task force, per year. Debbie Aquino manages the "Stand Tall, Don't Fall" and BADA Tobacco Grants, with a focus on integrating the "environmental strategies" component of the Communities That Care prevention model with use of the Developmental Assets framework. Aquino also coordinates promotion and communication for the coalition, such as managing HCC's Web site.

Deborah Loesch-Griffin is a community development and evaluation consultant whose consulting business, Turning Point, Inc., has been used by the coalition staff to develop assessment tools for their activities, and to assist with meeting grant accountability criteria. She is cofounder, along with her husband, Sean Loesch-Griffin, of Community Chest, Inc. Loesch-Griffin is an expert in the field of service-learning and has been instrumental in promoting the Developmental Assets as congruent with service-learning. Michelle Watkins has been the director of Central Lyon Youth Connections since 1990. She has completed an MSW thesis titled "Youth as Partners in Community Decision-Making." Prior to HCC's becoming a nonprofit organization in January 2002, Central Lyon Youth Connections was the fiscal agent for the coalition. In 1993, Watkins and Loesch-Griffin partnered to apply for a prevention grant. This partnership, which linked Lyon and Storey counties, became the organizational foundation upon which HCC was built several years later.

Linda Lang is a specialist in rural coalition building and nonprofit management who has had a relationship with the community leaders over the past 10 years. It was the interview with Lang that helped us understand the history of "coalitions" in northern Nevada. She worked for the state Bureau of Alcohol and Drug Abuse for several years, and is invaluable as a consultant on work with that organization as well as on the management of the new nonprofit status. When we made our second visit to HCC in June 2003, Lang had begun working with the Fernley task force. Fernley is one of the coalition communities that has experienced a recent population boom. As the population has increased, so has the demand for prevention, intervention, and youth development resources within the community.

Sector Representation

In addition to interviewing staff, we explored sector representation by reviewing print materials such as attendance rosters from coalition meetings and progress reports to funding agencies. According to the print materials, local, state, and county government agencies have been particularly active in the coalition, including Parks and Recreation, the Division of Aging, State Welfare, Public Health, and the sheriff's office. The Cooperative Extension Department of the University of Nevada is an instrumental sector involved with the work, particularly in the area of service-learning in the schools. Youth-serving organizations such as the Boys & Girls Clubs, 4-H Clubs, Little League, libraries, Central Lyon Youth Connection, and juvenile court actively participate in the coalition. Larger family service organizations, such as Community Chest, Family-to-Family, Red Cross, and visiting nurse services, are represented consistently on meeting rosters. Ethnic minority representation includes Tribal Social Services and the Hispanic Council. There have been challenges to engaging the media and business sectors. As for the media sector, there is simply little of it in the northern rural areas. As one focus group participant explained:

> It's hard to get the word out up here ... Twenty percent or under get the newspaper ... We have no local radio or ... TV. And it's kind of interesting. Sometimes we'll go to Reno and try to get, you know, the TV station to come out here, and they will not come out unless it's something bad ... [Or] they'll come down for the opening of a new audio store or something, you know, in Carson City. But not ... for community events.

Since the media are businesses dependent upon advertising sponsorship, radio and television in Reno and Carson City, for example, would need to convince advertisers that rural listeners and viewers have ample purchasing potential. There would seem to be little likelihood of succeeding in this effort. Despite the fact that there is not a large business sector in the coalition area, and there is concern that the same businesses are asked repeatedly to support the coalition, HCC has had some success in engaging the business sector. In addition to monetary and in-kind donations, businesses have been involved in the Solid Ground program in which youth volunteers are placed in work settings to learn job-related skills, such as appropriate attire and language, consistent attendance, and follow-through on assignments.

Casinos and brothels are two of the businesses that have been supportive of the coalition. In fact, one staff member said the Boys & Girls Club would not survive without their financial support. Although this support may meet with some ambivalence, given the mission of HCC, many parents are employees of the casinos, which are a crucial part of the state and local economy. Although we did not succeed in our attempt to interview a brothel owner, a drive down the access road to "The Bunny Ranch" took us past a child care center. The juxtaposition of the child care center to the brothel parking lot reminded us that many employees of brothels are also parents in the community, and should not be overlooked in HCC's work.

Indicative of their conviction in grassroots leadership and representation, the directors insisted that we visit all six communities during our visits. Having had this experience, we came away with a much deeper appreciation for the work involved in building and sustaining a rural network over several hundred miles. In the near absence of public transportation and local media, face-to-face communication and networking are critically important to raising awareness and supporting mobilization on behalf of Developmental Assets. Another aspect of the work, underscored by driving the distances between towns, is the professional isolation from colleagues.

The spontaneous conversations, meetings, coffee breaks, and lunches that occur in urban agency offices are not a part of the working environment, and since its beginnings, the women of HCC have been front-runners in recognizing the importance of building in professional support for the work.

Prevention and the HCC History of Coalition Building

In 1993, Michelle Watkins, director of Central Lyon Youth Connections in Lyon County, partnered with Deborah Loesch-Griffin, cofounder of Community Chest, Inc., in Storey County to apply for a state prevention grant. Watkins recalled the following:

> We didn't have the expertise to write grants. And Deborah Loesch-Griffin did up in Virginia. And, so, we wrote the first BADA one together, and we got it, to do prevention. And then from there it just kind of snowballed, because Debby's very, very good at looking for resources. And, so she found dollars from AmeriCorps to start in our communities. And we saw kind of a natural partnership. The Storey County line comes right down to Lyon County, and so we're serving some of the same individuals ... Plus Storey sometimes is placed off by itself, or they try to incorporate it, maybe, with a bigger town and it just doesn't fit for them. So, it seems like we have a lot more in common with Storey ... and it's pretty spread out too. So, that's how the partnership kind of began with that.

As successful recipients of the Bureau of Alcohol and Drug Abuse grant, the two communities were partners in a statewide prevention coalition prior to becoming HCC. BADA is one of many federal and state agencies that has adopted the Communities That Care (CTC) and community change model, and prescribes its use with grantees. Use of the CTC model preceded the coalition's awareness of Search Institute's strength-based approach to youth development through promotion of Developmental Assets. Based in prevention science, communities using the CTC model identify the risk factors inherent in their environment that could lead to high-risk behaviors in young people, such as substance use, violence, school dropout, and risky sexual practices. Communities also identify the protective factors that foster positive connections to supportive adults and help prevent children from engaging in high-risk behavior. CTC offers a fully developed "operating system" that provides communities with a detailed handbook for taking systematic steps in assessing risk factors and community resources, and for creating and implementing an action plan. Implementing the action plan involves suggestions and guidance for educating the community, building collaborative relationships, and obtaining resources. The CTC model is attractive to funders because it has extensively validated evaluation tools to assess outcomes:

> When we started the asset development it was really with the Drug-Free Communities dollars ... And then we have our state agencies coming along and saying, "Well, we want to help build coalitions, also. Here are some dollars available." And so, we were like, "Oh, OK!" So we wrote for these dollars ... And we had no idea, when we wrote it, we had no idea of what kind of model was going to be used. And it's totally [focused on] risk and [protection], so that became a challenge right from the beginning, we started going to the

> trainings ... That's when I first realized that the [Communities That Care] model was not like Developmental Assets. (Leadership focus group participant)

In the leadership focus group, coalition members described many positive aspects of having begun their formal organization with the CTC model. The HCC Web site summarizes aspects of the model as follows:

> Prevention science is a means of acquiring and understanding information about risk and protective factors present in a young person's developmental environment and putting those findings to work in achieving desirable behaviors. [J. David] Hawkins and [Richard F.] Catalano [CTC's founders] have developed a strategy called the Social Development Strategy that organizes this research into a framework of community-supported actions that lead to the optimum environment for positive youth development.

Similar to the Developmental Assets framework, CTC is theoretically based on the importance of human relationships and environmental supports. Healthy bonding throughout childhood is viewed as the key to lifelong healthy development. The premise is that the human predilection for bonding reinforces the adoption of group standards. For example, youth will be more inclined to accept the standards of violent gang behavior in the absence of positive alternative affiliations such as a youth civic leadership group. In order to decrease risk behavior, the CTC model promotes the community's development of opportunities for youth involvement, skill building to ensure successful participation, and reinforcement for skillful performance. The model further stresses that providing opportunities for youth is not sufficient. The community must provide the skills training youth need to access and have positive experiences of these opportunities. In this model, youth training includes cognitive skills, such as problem solving and decision making, and functional skills, such as reading, writing, and mathematical competence. The CTC model also cautions against adults' tendency to focus on negative youth behaviors instead of positive ones, which is consistent with its emphasis on positive reinforcement rather than punishment.

Clearly, there are many conceptual similarities between CTC and the Developmental Assets framework. From a utilization perspective, the CTC model has developed a product line of surveys and implementation guides that community groups can purchase. Search Institute also offers materials for purchase that include surveys and training guides for supporting implementation of a Developmental Assets approach to positive youth development. The focus group with HCC leaders provided a retrospective comparison of how each of the models had served the coalition. In the next section, we present excerpts from the leadership focus group transcript in order to let those involved with HCC address the strengths and weaknesses of the CTC model from their own experience. We present this material to illustrate how HCC's resourcefulness, creativity, and determination have significantly contributed to the notion of a *blended* model of community change that can incorporate the best of both CTC and Search Institute's Developmental Assets framework.

Providing the Map

The Communities That Care model clearly facilitated HCC's movement from an informal network of service providers to an organizing force in the community.

> The coalition was basically, you know, just partners that met, and we all kind of took on roles. And the county took the role, you know, a person there helped us do minutes and agendas and that kind of thing. And now we're moving toward formalization. And with the Drug-Free Communities and the Coalition dollars for BADA, we hired a Coalition Coordinator ... [CTC] really laid a nice foundation for our coalition becoming more formal ... it was a very natural step. (Leadership focus group member)

A first step in the CTC model is community mapping, a process in which communities take inventory of their programmatic and service resources. This step enables them to identify services and update community needs. This step was particularly important for the northern Nevada communities because they were undergoing significant population growth:

> The community mapping is awesome. We need to do it. Our communities, some of them are growing rapidly. We don't even know what we have as far as what kinds of programs we should be looking at. Where are the gaps in service? (Leadership focus group member)

Further, the CTC model provided clear, concrete strategies for how to use the information gained in the needs assessment. One of the codirectors said, "They hear 'needs assessment' ... and a reaction I've gotten many times is 'Not again. We have 15 on the shelf.' But Communities That Care does do a good job in kind of bringing you past just putting it on the shelf."

Early Awareness and the Appeal of Developmental Assets

When HCC became aware of Search Institute's strength-based orientation to healthy youth development, they were encountering challenges in the use of the CTC model with their Task Forces. A codirector recalled the following:

> The CTC model does not work in the task forces. We tried to do that. CTC does not fly at the task force level. Because, I mean ... do you guys care about logic models? No, they want to hear about assets and how they can build. And, so that's why we've mixed the two.
>
> ...
>
> These communities here, they are very grassroots. That's how they have worked for 10 years. And then all of a sudden we have this CTC model that's very formal ... And so there was resistance on, you know, from our community members and some of our coalition members as well. You know, [their attitude is] "Why change it? It's working okay now." ... CTC is basically a logic model where you have to assess your needs ... Where we've run into problems right now is that they really are forcing use of these best practices in prepackaged programs that are very expensive, very time-consuming. And, you know, we don't have the staff out here. We don't have the qualified people.

The formality that had helped to effectively organize the coalition was not adaptive for the task forces. No individual community had the financial or staff resources for implementation. The emphasis on professional resources and packaged guides was inconsistent with a grassroots community initiative that focused on the expert wisdom of the people who had been living in the community. A leadership focus group participant said the following:

> You're supposed to have this big bank of programs you can choose from ... And it's like ... thousands of dollars for these programs, and then full-time personnel, and you know, full integration with the schools. And our schools are not ready to do that. And yet they say, you know, "Well, if you want to still continue to get state dollars you must do this."

The Search Institute model is based on the premise that the local community is best positioned to determine how to apply the Developmental Assets. Although the lack of a standard community model is a criticism often leveled at the institute, the adaptability encouraged by the HC • HY initiative works well for these small, rural communities. The list of internal and external assets was found to be highly accessible and empowering of the everyday people who translate the theory into action. Further, after much emphasis on the negative aspects of living in rural communities—the isolation, lack of convenient shopping, lack of human services, and poverty—there is a strong appeal to focusing a positive lens on the community and its youth:

> I mean how many of us can stand up and say everything that's wrong in our community? We see it every day. We see it in the news. And so, when you turn around and look at assets, it's a different approach. It's a positive approach. And people seem to understand it better too. (Staff focus group participant)

The strength-based orientation and accessibility of the Developmental Assets is also appealing to youth who see themselves presented positively in the materials. That has not been HCC's experience with Communities That Care. As a staff member put it: "It's hard to engage the youth with 'Communities That Care.' There's just nothing attractive about it [for them due to its focus on risk factors]. I mean, even with adults, it's hard."

Development of the Blended Model

The Lawton/Fort Sill Community Coalition (Chapter 8) and HCC have contributed extensively to our understanding of the feasibility of using a blended approach to community change on behalf of child and youth development. Both coalitions have multiple prevention-based funding sources, which require specific evaluation of prevention outcomes. Both communities also face heightened risks to the well-being of their youth. In Lawton/Fort Sill, the presence of a large military base has been associated with high levels of substance use and precocious sexual behavior among the area's youth. In northern Nevada, legalized gambling and prostitution are among the environmental risk factors that pose particular challenges. It is not surprising, then, that the impetus for blended prevention and Developmental Assets models have come from these community initiatives.

Developmental Assets and Protective Factors

The Healthy Communities Coalition has established its own "fit " for the Developmental Assets within its prevention programming by expanding the notion of protective factors:

> And then there [are] the protective factors. But they [CTC and the organizations promoting prevention approaches] don't tell you how you can measure them, and they don't tell you how you can identify them in your community, how you can promote them. They tell you the theory that, if you promote them, it helps reduce the risk. But that's where I think a lot of communities are trying to figure out [how to], you know, plug in asset building because it's good for the community. I mean, it gets them engaged. It gets them involved. (Task force focus group member)

Drawing from prevention science, the CTC model states that protective factors buffer children against high-risk behaviors. Protective factors are described as "individual characteristics," "bonding," and "clear standards and healthy beliefs." HCC and other initiatives are discovering that such definitions of protective factors are not explicit or practical enough for use by the average community member who is not professionally trained in youth work. Further, these descriptions overlap with many of the Developmental Asset categories or clusters, such as Boundaries and Expectations, Commitment to Learning, Positive Values, and Social Competencies. The practical advantage of these asset categories, however, is that they are further defined by the simple actions or experiences (the 40 assets) that comprise these larger conceptual categories. The HCC codirectors have been framing the specific assets as protective factors in their work, and introducing the language to state funders. One of the codirectors said the following:

> I was hoping this might be our crack in the door to some of our state funders, with these environmental strategies, wow, you know, this sounds a lot like assets, and they may be able to start looking at the assets as worthwhile stuff.

After some resistance, the codirectors began to witness the shift they were looking for:

> There was a point where you almost tried your hardest not to mention it to the state agency, that you were using Developmental Assets, I mean, or else you were in big trouble . . . Back in November [2002] we had a meeting—and the state agency was there—and a lot of people were saying, "Oh and we're doing Developmental Assets." And I thought, You know . . . they're not going to say no to 20 communities saying that.

In describing HCC, its history, and its conception and implementation of a blended model for promoting youth development, it warrants reiterating an achievement that is often buried in assessments of impact, an achievement that in and of itself is an act of community change. In a large coalition such as HCC, which serves disparate communities separated by geography and their own local concerns, establishing an effective communication infrastructure is a critical achievement. HCC has fostered connection of each partnering community to its respective task force, as well as having linked the six community task forces to the larger coalition, and the larger northern coalition to the Nevada statewide initiative. The very development of this communication network allows the partnering communities to strengthen one another and better provide the asset-based activities generated by the coalition. In other words, part of HCC's blended model includes the provision of broader community support for local youth and adults who are carrying out the asset-based and prevention activities. Lack of community resources is a risk factor; making asset-based resources more readily available through the coalition network is a protective factor, one that HCC is providing on a daily basis.

A crucial part of HCC's communication infrastructure has been the development of an evaluation protocol for assessing impact across the coalition. HCC has devised a system for gathering highly detailed information as part of their prevention grant accountability, and continues to create tools for capturing positive impacts beyond risk reduction. Developing this coalition-wide system has led to close examination and comparison of prevention-oriented and strength-based community change models. In blending the application of these models, HCC has shown that risks and strengths do not have to be framed as polar opposites, and that each orientation has a contribution to make to community coalition building in the service of healthy youth development.

Applying and Assessing the Blended Model

The initial focus of HCC's activities emerged in response to the assessment of risk factors identified in each of the communities through the CTC needs assessment. These risk factors include particular attitudes and behaviors of adults and youth as well as a lack of certain environmental resources. HCC's findings concerning alcohol abuse, for example, can be viewed as a constellation of adult and youth attitudes, perception, and behaviors, which place youth at particularly high risk for alcohol use. Adults were found to be generally accepting of underage drinking, and youth were found to have an inflated perception of their peers' alcohol consumption. This combination of attitudes and perception places youth at heightened risk for their own alcohol and abuse. A significant focus of HCC's activities has been to enhance protective factors against such risks, through increasing prevention programming in each of the six participating communities.

From the perspective of prevention theory, learning about the risks associated with specific behaviors provides a certain degree of protection against those risks. A second step in most prevention models is to educate the community on the risks associated with the targeted behaviors and on the protective factors that can be promoted to reduce the risks. HCC, working from the CTC and other prevention models, utilizes youth, family, and larger community prevention strategies. In addition, however, coalition members build Developmental Assets approaches into their orientation toward protective factors. From a Developmental Assets perspective, the greatest protection against negative behaviors, such as underage drinking, is active engagement in positive behaviors with prosocial peers and caring adults. HCC relies on basic strategies for encouraging such behaviors, including the involvement of youth in the community task forces and engaging as many youth as possible in service-learning projects in which they work with peers and supportive adults to improve their community.

In its efforts to enhance protective factors to buffer youth against high-risk attitudes and behaviors, HCC strives to establish and maintain two to three youth programs at all times through each of its six task forces. They have developed these programs in collaboration with community organizations, including Central Lyon Youth Connection, 4-H, Boys & Girls Clubs, the schools in Lyon and Storey counties, a service-learning organization, and the Lyon County Parks and Recreation Department. The programs that have been established have targeted before- and after-school time and summer hours, addressing the assets in the Constructive Use of Time category. The content areas have been in community service, mentoring, homework completion, and prevention of early substance use.

To some degree, most of the programming uses youth-to-youth mentoring in which older students serve as role models to younger ones. Mentoring is also a service-learning activity for which older youth receive academic credit. High school students connect with younger students as reading buddies and through other established programming such as 4-H. High school students in Yerington, for example, performed theatrical skits on drug-free lifestyles for elementary school students. HCC has also developed adult mentoring programs for youth, as a way of promoting youth relationships with positive adults outside their families. Earlier in this chapter, we referenced one such program, Solid Ground, which has a particular focus on mentoring in the area of work habits and career development. Another mentoring program, in Dayton, matched youth involved in the juvenile justice system with adults. A program in Virginia City organizes field trips for 5–12th graders as positive alternatives to drug and alcohol use. The field trips include peer- and adult-facilitated trainings in leadership development and conflict resolution. All of this programming features positive activities with peers and adults, which gets to the heart of HCC's blended model: Even when focusing on prevention activities like tutoring to address academic risks, substance abuse awareness, and conflict resolution training, the goal is to feature caring and supportive relationships, thereby building the assets in the Support category, minimally, but typically in one or more of the other asset categories as well.

HCC primarily relies on prevention and service-learning grants to fund the youth programming, and several prevention-oriented curricula are interwoven within mentoring, homework, field trips, and after-school activities. Stand Tall, Don't Fall is a curriculum that utilizes a youth advocacy group to glamorize youth sobriety and abstinence from alcohol use until age 21. Teens Against Tobacco Use and Why Am I Tempted? are other examples of the prevention content included in youth programming. Over time, service-learning has been a primary vehicle for incorporating the strength-based orientation. In the following sentence, Developmental Assets that are addressed through specific components of service-learning activities are named in parentheses: inclusion of student voice (Youth as Resources), high-quality community service opportunities (Youth Programs; Service to Others), and adult- and peer-supported reflection on strategies (Adult Role Models; Positive Peer

Influence) and accomplishments (Self-Esteem). It is specific assets such as these that community participants comprehend as providing protection against risk. The asset language and activities concretize the notion of protective factors, while also positively transforming the language used to describe the approach.

Does It Matter? How Do We Know the Blended Model Works?

The blending of prevention and Developmental Assets approaches requires a blending of assessment approaches as well. Before we move into the overview of formal evaluation activities associated with the prevention side of the equation, it is critical to note the qualitative impact we could detect through our interviews and observations.

First, the Empowerment assets were featured in virtually every activity we either heard about or observed. HCC stands out for its emphasis on supporting youth voice in its task forces, general activity planning, and in leading activities. This initiative was as youth-centered at the leadership level as any we studied. The involvement of substantial numbers of youth in leadership activities is both an asset-based process and outcome. HCC is having a clear impact in the construction of empowering activities and in its creation of opportunities that fall within the Constructive Use of Time asset category. We cite these as examples of how impact would be assessed from a Developmental Assets perspective, not as a definitive summary of where the impact is greatest. HCC's evaluation system tracks youth involvement thoroughly across the coalition and uses those findings to support its accountability claims.

Is the blended model having an impact on the internal asset categories, such as Positive Identity and Social Competencies? Those data were not available at the time of our study, and they will be important to collect as the blended model evolves. The key question here is whether the cultivation of the external assets translates into the growth of internal assets. Ironically, HCC is better positioned to measure the impact of the external assets on the reduction of risk, given the coaliton's employment of the prevention-mandated assessment tools.

Turning Point, Inc., the consulting group cited earlier in the description of staffing and leadership roles, has played a key part in developing evaluation tools and methods of analysis for the assessment of HCC's work. Systematic evaluation requires addressing the difficulties involved in measuring impact in very different communities that are often unevenly represented in response rates to survey instrumentation. "The Millennium Report for Change: A Status Report on the Risks and Assets of Five Communities in Storey and Lyon Counties," profiles each community task force's challenges, priorities, and goals. Identifying and prioritizing risk factors through community mapping of resources and risk factors is part of the CTC methodology mentioned earlier as a useful initial step in assessment, and has been a major achievement of HCC. Of the wide range of evaluation strategies used to determine the impact of the coalition's activities, examples include assessing the extent to which youth are making healthy choices before and after participating in prevention curricula. There are measures of family behavior regarding substance use and whether parents have talked with their children about alcohol, tobacco, and other drugs. Another example is surveying or debriefing youth about the quality of their experiences as members of the task forces working with adults.

The coalition uses surveys popular with funding agents, such as BADA and OJJDP, and correlates results from these instruments with other sources of data. For example, in assessing youth alcohol and tobacco use, *The Youth Risk Behavior* survey is administered, along with tracking tax revenue generated from liquor license sales and taxes on alcohol and cigarettes. Records of juvenile arrests are examined for incidents of property destruction versus driving under the influence of drugs or alcohol. Education databases provide a source of information concerning school dropout, teen pregnancy, standardized test scores, and attendance records. The *Partnership Assessment* survey has been developed to query businesses, agencies, and organizations about the extent to which coalition activities are perceived as competently led and the accessibility of information about coalition activities. The *Civic Responsibility* survey measures the impact of service-learning, a central vehicle for operationalizing Developmental Assets in northern Nevada.

Voter turnout rates and numbers of building-permit applications provide a sense of the level of community connectedness. In addition to surveys and tracking of demographics, the evaluation infrastructure has included interviews with key participants and short reflection surveys with open-ended responses to capture youth experiences of service-learning projects. Establishing and managing a complex evaluation infrastructure of this magnitude throughout two counties is an impressive accomplishment. It requires highly labor-intensive activities that are often met with resistance by service providers and community members—adults and youth—who prefer "doing the work." Further, the evaluation efforts may not show hoped for changes in youth risk behaviors in the short term. In fact, at the time of our study, there was much more information about what was being assessed and how it was being measured than there were concrete positive results, with the clear and critical exception of youth involvement in the broad range of activities presented here.

The New Frontier

The Healthy Communities Coalition of Lyon and Storey counties should inspire a great deal of hope for community change agents who want to exercise their creativity and retain their collaborative spirit. Two thematic clusters from our larger case study analysis have not been featured in the prior community profile chapters, and they apply in interesting ways here: *Resisting the Mission* and *Does It Matter?*

Resisting the Mission is a cluster of codes from our analysis that depicts individual and sector-specific resistance to the HC • HY mission of pursuing community transformation on behalf of strength-based youth development. Such resistance is typically a result of inadequate evidence in support of the mission's targeted goals, a desire to hold on to the usual ways of doing things, or an inability to commit to the mission's work for other reasons. HCC presents a dramatic variation on this theme. It found itself partially resisting the larger prevention mission that evolved over the course of its work in response to available prevention funding and dominant prevention models. As HCC discovered and introduced the Developmental Assets framework to state funding agencies, its efforts were initially met with a great deal of resistance. In fact, from a financial perspective, the resistance was too thorough to completely overcome. HCC's response to resistance from the funding sector was to create an integrated model for supporting youth that met the funders' needs and expectations while simultaneously advancing its own emerging commitment to strength-based approaches.

At the time of our study, HCC's blended model of developmental assets and prevention work might best be termed *Revising the Mission*. The coalition retained a commitment to certain aspects of its prevention roots. It embraced the community-mapping exercises that focus on community needs and resources, and it conceptually embraced the notion of protective factors central to prevention science. Over time HCC progressively infused more elements of the Developmental Assets orientation into its prevention activities. By concretizing the protective factors in terms of specific Developmental Assets, and building those assets through combinations of mentoring and service-learning projects, HCC was moving comfortably into a new frontier in this work, the frontier of blended models. There may have been resistance to both a pure prevention approach and a pure strength-based approach in their emergent orientation, but their work did not smack of resistance; rather, the tone was one of resonance. It seemed the coalition had struck the balance it found right for its community at this point in time.

Does It Matter? is the thematic cluster intended to capture the myriad ways in which initiatives assess the value of their work. HCC has much to offer on at least two scores with respect to this issue. First and foremost, from our perspective, "mattering" should be assessed based on the meaning of the work for the youth involved. By creating its six-community coalition in northern Nevada, HCC developed a vast array of youth programs and activities, with many of them organized by youth and featuring service to others. We saw more authentic youth involvement in HCC than perhaps any of the other seven initiatives in our study. The introduction of the Developmental Assets approach to the six community task forces proved pivotal to cultivating youth interest and involvement. As the coalition's codirectors reported, it was difficult to elicit authentic engagement from youth with regard to the prevention curricula only. The language was too abstract for them, and the issues too problem focused. Bringing asset-building activities into the planning groups seemed to wake up the youth, and served to functionally turn co-ownership of the coalition over to them. The coalition came to matter to youth when assets became the focus.

The second level on which HCC is making a critical contribution to the national HC • HY initiative is through its systematic and multifaceted efforts to assess both prevention and strength-based outcomes. Consistent with its blended model orientation, HCC is carefully tracking levels of risky behavior, protective factors, and now positive behaviors, relationships, and community development activities. By partnering with evaluation specialists in northern Nevada and sharing findings across the larger state prevention and Developmental Asset networks, HCC is breaking new ground on the evaluation front as well. If the coalition's assessment efforts ultimately shed further light on the interrelationships between prevention and asset outcomes, the next new frontier may again be discovered in northern Nevada.

Postscript

The Healthy Communities Coalition of Lyon and Storey Counties works with youth leadership teams in all of the area's eight high schools and seven middle schools to connect adults and youth in the communities and bring the Developmental Assets into schools.

The youth leadership teams come together at a summer camp to build leadership skills and share ideas as they plan how they will promote the assets in their schools over the next year.

Adult representatives from various community sectors present information to the youth at the camp, and then work under youth leadership to integrate the information into their plans for the school year.

The yearlong plans include at least two asset activities per month. After the plans are created, the leadership teams meet with their principals to have their plans approved. Some of the accomplishments of the youth leadership teams include the school district's approval of community service as an activity in which students can earn a letter, and the school board's decision that the completion of 120 hours of community service is equivalent to earning one-half credit. This development has engaged credit-deficient students, allowing the Healthy Communities Coalition opportunities to work with leaders who often have not self-identified as such.

Chapter 11
Project Postscript: Resisting the Template

Despite the call for "definitive guidelines" for implementing a Developmental Assets approach in one's community, the eight interpretations of the framework presented here portray the value of communities working out approaches that fit best within their particular contexts. In the last of the community chapters, we learned of northern Nevada's struggles with a more standardized prevention-based community change model. In the case of that local initiative, the freedom to adapt the Developmental Assets framework was critically important, including the freedom to blend it with other approaches. This is not to say that "best practices" should not be developed and shared, or that there should not be standards of effective implementation; we argue, however, that there is more to gain by continuing to examine the diversity of community youth development approaches emerging from the field than by offering definitive and comprehensive implementation models. In this concluding postscript, we present features of context-specific effective practice, including those that go "sector deep," as well as those that exemplify effective cross-sector collaboration. This conclusion is not intended to be a summary of the overall study, however, as the comprehensive findings have already been presented in Chapter 2, with individualized interpretations presented in each of the eight community chapters. Rather, this postscript underscores some key themes that others interested in pursuing a Healthy Communities • Healthy Youth (HC • HY) initiative will want to consider as they move forward.

Where to Start?

One of the most fascinating things we learned in our study of the eight communities and their HC • HY initiatives was the manner in which they began the work and how that influenced subsequent decisions. Interestingly, most of the initiatives chose to pursue an HC • HY approach in their communities in response to dissatisfaction with the limitations of existing work they had been doing, prevention, intervention, or some other form of youth-service work. These communities often shared Search Institute's view that most of the existing approaches either implicitly or explicitly worked from deficit-based perspectives on youth. In this regard, their starting points

M.J. Nakkula et al., *Building Healthy Communities for Positive Youth Development*, The Search Institute Series on Developmentally Attentive Community and Society 7, DOI 10.1007/978-1-4419-5744-3_11, © Springer Science+Business Media, LLC 2010

were some combination of dissatisfaction with the results from existing ways of supporting youth and a philosophical orientation toward youth development. Second, the communities that choose to become HC • HY initiatives are seeking ways to involve the larger community in youth development efforts. The "professionalism" or "paraprofessionalism" associated with formal prevention and intervention programming, for example, rules out or is perceived to rule out the average person from participating in the work. The HC • HY approach is intended to fundamentally change that reality; at its core, it is a community change effort on behalf of youth development rather than a program or initiative designed to promote it.

Although there were similar philosophical reasons for beginning the work across the eight communities, these philosophies were put into practice in very different ways. Typically, there were one or two focal community organizations or programs that served as the initial host of the effort, with each having one or two champions who rallied the community around the work. One of the most common themes from our study was the dynamic relationship between determined leadership and the everyday participation of the broader community. McPherson, Kansas, provides a clear example of the role of central organizations and key determined leadership. The McPherson asset-building initiative was deeply rooted in that city's Chamber of Commerce, with the Chamber vice president serving as the de facto initiative coordinator. Because a community's Chamber of Commerce is charged with reaching out to the larger community, particularly as that applies to connections between the civic and economic health of the community, that sector holds specific appeal from an asset-building perspective. However, community-wide asset building requires a great deal of commitment, and coordinating such efforts out of a Chamber of Commerce requires a special commitment. McPherson had that commitment in its vice president, who was determined to support the development of leadership skills for all residents—adults and youth—as a means of strengthening the community. McPherson's key partner was an outside organization—the Kansas Health Foundation—which shared a passion for leadership training and its relationship to community change. The synergy developed between the local Chamber of Commerce and a powerful state nonprofit organization that provided funding gave the McPherson asset-building initiative a coherent center and a strong springboard for broader community-building efforts.

The Traverse Bay Area initiative in northern Michigan is a classic example of locating the initial work primarily in one sector at the beginning, developing it there, and moving out strategically. This community began its work in the public schools and did so by recruiting all 19 superintendents from the area. Getting buy-in at the top allowed the initiative to gain early momentum in that sector and build from there. Traverse Bay also had a key funder involved as an early partner—the United Way—which was critical to building integrated funding across partnering organizations.

Northern Nevada's HC • HY initiative was rooted in prevention programming. In its efforts to link six rural communities through prevention-based activities that responded to risks associated with rural poverty and isolation, and the state's legalized gambling and prostitution industries, the leadership team for that prevention work had been struggling to obtain stronger youth participation. The Developmental Assets approach, with its positive orientation, proved to be an effective recruitment vehicle for youth as well as adults. As a result, this initiative has constructed innovative approaches to blending the asset and prevention-based work, and in doing so has enhanced youth involvement and stimulated broader community participation in youth-development activities. If the Nevada initiative's intensely determined leadership team had not begun its work in the prevention arena, and had it not been so heavily reliant on prevention funding for its sustainability, it likely would not have developed the blended approach to the work that now defines its identity. Where and how an initiative starts the work certainly influence how it moves forward, but many other factors contribute to the evolution of its efforts as well.

Key Bridges

Because of the emphasis on involving the whole community in the commitment to promoting healthy youth development, key people or institutions become critical to functionally linking one sector to another in the service of the work. We came to see those connectors as *key bridges* over the course of the study. Orlando targeted "White guys over 50" as instrumental to their work, precisely because successful businessmen and other professionals not typically associated with youth development were viewed as essential to the initiative's truly becoming a community-wide entity. By reaching particular men who fit this description, the Orlando initiative was able to cultivate a cohort of ambassadors who could reach out to other powerful leaders in the community. Virtually every one of the initiatives identified key bridges to cultivate, and those key bridges, in turn, cultivated others.

Moorhead, Minnesota's initiative, was exemplary of a common approach among HC • HY initiatives in this study and beyond: Moorhead targeted leadership in the community's police force as a key bridge. This tactic serves multiple purposes. First, it serves to create the recognition of the larger police department in the work, and in doing so introduces the philosophy of positive youth development to that critical institution within any community. In some cases, the police chief promotes asset training among police officers, and in many cases, a strong partnership with the police department provides important bridges to teen centers, schools, parks, and other areas where the police have a strong presence. In the case of Moorhead and elsewhere, bridging the police department and the local schools as a central early cross-sector partnership provided a powerful model for additional intersector collaborations.

In many cases, youth themselves are key bridges in the initiatives. As is common knowledge, youth with particular types of social status hold tremendous influence in the lives of their peers. Two dominant types of youth-specific key bridges emerged from the study. The first is the youth leader who seems to be involved in almost everything sponsored by the community, whether in or out of school. These youth are effective in reaching out to others because of their energy and intelligence and sheer enthusiasm for being at the center of things. They tend to link with other influential youth in organizing school- and community-based projects that give the initiatives another level of visibility. There is a second type of youth-specific key bridge, however, who also is critical to building the initiative; it is the young person who is particularly adept at connecting with marginalized subgroups of youth. Linking with the *skaters* who might intimidate many adults with what can appear to be a countercultural orientation in dress, body piercing, tattooing, and language, youth who are either insiders of this group and connected to the initiative, or those who are not insiders but able to affiliate with such groups comfortably, have an extraordinary role to play in the HC • HY movement. Such "crossover kids" afford young people who see themselves as outsiders a better chance of finding a role in a community-wide effort. In fact, we would argue that cultivating this second type of youth-specific key bridge is essential to the broad-based success of each local initiative, and ultimately to the national HC • HY movement.

Idiosyncrasies and Blended Models

If the eight community interpretations presented here teach us one core lesson to hang our hats on, it is that the HC • HY approach must be tailored to community-specific needs, interests, and strengths. The passion that inspired Portland to make the media a central focus in its work should be reinforced, as it clearly led to positive breakthroughs in that community, but it would not seem wise to prescribe a media focus as a general rule any more than it would seem wise to reinforce relationship building with the police, the business sector, or the public schools. These are all critical sectors to community-wide collaboration, and it seems clear that communities should begin in their areas of strength and build from there. Nonetheless, all HC • HY initiatives have much to learn from those communities that have cultivated particular sector-specific strengths. They also have much to learn from community-specific approaches to cross-sector collaboration. Not everyone will want to target the "White guys over 50" any more than they will want to target "the usual suspects" who do all the work, as many initiative leaders described it. Not every community will be able to cultivate the powerful relationship with a state-funding organization in the way McPherson, Kansas, did, yet there's much to learn from their model.

Toward the end of our study, it was becoming clearer that some HC • HY initiatives were developing blended approaches with prevention partners. This decision evolved out of a combination of funding realities, community needs, and philosophical orientation. Many states and foundations invest their youth development money most thoroughly in the prevention work and its focus on quantitative evaluation. The HC • HY initiatives in Lawton, Oklahoma, and northern Nevada, in particular, were reliant on prevention funding for a good deal of their asset work and were philosophically supportive of prevention activities for youth in their communities. At the same time, they embraced the need for a community-wide strength-based orientation. Those two initiatives have made great strides toward integrating the two approaches and are philosophically comfortable with that position. At this point, they do not present blended models for replication in other communities; rather, they provide examples of blended approaches that work for them, in their particular contexts.

We are not concluding that blended-model approaches point to the next wave of asset-building work. The St. Louis Park initiative, for example, points out many of the potential pitfalls of becoming reliant on prevention or other external monies. They view the initiative as community work and argue that becoming beholden to the funding structures of standardized approaches moves one toward a slippery slope of regressive thinking. We do not take sides on this issue, given what we have learned from the eight communities. We defer to the local knowledge of the community leaders, which, as the initiative in McPherson would argue, should ultimately be the knowledge of the whole community of leaders. These eight interpretations are most useful for their diversity of offerings, rather than for any standards that might be inferred from their challenges and successes. Our hope in that vein is that the interpretations presented here inspire further innovation rather than imitation. If that hope is realized, so too will be the hopes of the children such work represent.

Bibliography

Benson, P. L. (2006). *All kids are our kids: What communities must do to raise caring and responsible children and adolescents* (2nd ed.). San Francisco, CA: Jossey-Bass. (First edition published 1997).

Benson, P. L., & Pittman, K. (2001). Moving the youth development message: Turning a vague idea into a moral imperative. In P. L. Benson & K. Pittman (Eds.), *Trends in youth development: Visions, realities and challenges* (vii–xii). Boston, MA: Kluwer Academic Publishers.

Boyatzis, R. (1998). *Transforming qualitative information: Thematic analysis and code development.* Thousand Oaks, CA: Sage.

Brooks-Gunn, J., Duncan, G. J., & Aber, J. L. (Eds.). (1997). *Neighborhood poverty: Vol. 1. Context and consequences for children.* New York: Russell Sage Foundation.

Burt, M. R., Resnick, G., & Novick, E. R. (Eds.). (1998). *Building supportive communities for at-risk adolescents: It takes more than services.* Washington, DC: American Psychological Association.

Carnegie Council on Adolescent Development (1992). *A matter of time: Risk and opportunity in the nonschool hours.* Washington, DC: Author.

Comer, J. (1997). *Waiting for a miracle: Why schools can't solve our problems—and how we can.* New York: Dutton.

Feister, L., Reisner, E. R., Marzke, C., & Castle, A. M. (1999). *Executive summary of the preliminary report of beating the odds from 3–6 p.m.: Evaluation results from the TASC after-school program's first year.* Washington, DC: Policy Studies Associates.

Freedom Writers (with Gruwell, E.) (1999). *The Freedom Writers diary: How a teacher and 150 teens used writing to change themselves and the world around them.* New York: Broadway Books.

Gladwell, M. (2000). *The tipping point: How little things can make a big difference.* Boston: Little, Brown.

Gold, S., & Ellwood, D. (1994). *Spending and revenue for children's programs.* Washington, DC: The Finance Report.

Helge, D. (1989). *Preventing teenage pregnancies in rural America.* Bellingham, WA: National Rural and Small Schools Consortium.

Larson, R. (2000). Towards a psychology of positive youth development. *American Psychologist, 55*, 170–183.

Legacy 2000 (2000). Unpublished Report. Orlando, FL: Healthy Communities Initiative of Central Florida.

Mannes, M. (2002). Search Institute's evolving approach to community-based human development and the role of service learning. In M. E. Kenny, L. A. K. Simon, K. Kiley-Brabeck, & R. M. Lerner (Eds.), *Learning to serve: Promoting civil society through service learning.* Boston: Kluwer Academic Publishers.

M.J. Nakkula et al., *Building Healthy Communities for Positive Youth Development*, The Search Institute Series on Developmentally Attentive Community and Society 7, DOI 10.1007/978-1-4419-5744-3, © Springer Science+Business Media, LLC 2010

Mannes, M., Lewis, S., Hintz, N., Foster, K., & Nakkula, M. (2002). *Cultivating developmentally attentive communities: A report on the first wave of the national asset-building case study project.* Unpublished working report. Minneapolis, MN: Search Institute.

National Research Council and Institute of Medicine (2000). *From neurons to neighborhoods: The science of early child development.* Washington, DC: National Academy Press.

National Research Council and Institute of Medicine (2001). *Community programs to promote youth development.* Washington, DC: National Academy Press.

Newman, R. P., Smith, S. M., & Murphy, R. (2001). A matter of money: The cost and financing of youth development. In P. J. Benson & K. J. Pittman (Eds.), *Trends in youth development: Visions, realities and challenges* (pp. 91–134). Boston: Kluwer Academic Publishers.

Pittman, K., Irby, M., & Ferber, T. (2001). Unfinished business: Further reflections on a decade of promoting youth development. In P. L. Benson & K. J. Pittman (Eds.), *Trends in youth development: Visions, realities and challenges* (pp. 3–50). Boston: Kluwer Academic Publishers.

Posely, J. (1976). *The Park: A history of the city of St. Louis Park.* St. Louis Park, MN: City of St. Louis Park and Its Bicentennial Commission.

Putnam, R. (2001). *Bowling alone: The collapse and revival of American community.* New York: Simon and Schuster.

Ramsey, R. D. (1999). *The children first story: How one community partnership is learning to put its children first.* St. Louis Park, MN: Children First.

Resnick, M. D., Bearman, P. S., Blum, R. W., Bauman, K. E., Harris, K. M., Jones, J., et al. (1997). Protecting adolescents from harm: Findings from the national longitudinal study of adolescent health. *Journal of the American Medical Association, 278,* 671–694.

Roehlkepartain, E. C. (2001). *An asset approach to positive community change.* Minneapolis, MN: Search Institute.

Roehlkepartain, E. C., Benson, P. L., & Sesma, A. (2003). *Signs of putting children first: Developmental assets among youth in St. Louis Park, 1997–2001.* Minneapolis, MN: Search Institute.

Rogers, E. (2003). *Diffusion of innovations.* New York: Free Press.

Scales, P. C., & Leffert, N. (2005). *Developmental assets: A synthesis of the scientific research on adolescent development* (2nd ed.). Minneapolis, MN: Search Institute. (First edition published 1999).

Seligman, M. E. P., & Czikszentmihalyi, M. (2000). Positive psychology: An introduction. *American Psychologist, 55,* 5–14.

Strauss, A., & Corbin, J. (1990). *Basics of qualitative research: Grounded theory procedures and techniques.* London: Sage.

U.S. Bureau of the Census (2000). *Profile of general demographic characteristics, St. Louis Park City, MN.* Washington, DC: Author.

Vision St. Louis Park (1995). *A special report about our community's future.* St. Louis Park, MN: St. Louis Park School District.

Whalen, S. P., & Wynn, J. R. (1995). Enhancing primary services for youth through an infrastructure of social services. *Adolescent Research, 10,* 88–110.

Subject Index

D

E

MIX
Papier aus verantwortungsvollen Quellen
Paper from responsible sources
FSC® C105338

FSC
www.fsc.org

If you have any concerns about our products,
you can contact us on
ProductSafety@springernature.com

In case Publisher is established outside the EU,
the EU authorized representative is:
Springer Nature Customer Service Center GmbH
Europaplatz 3, 69115 Heidelberg, Germany

Printed by Libri Plureos GmbH
in Hamburg, Germany